Touring the Universe through Binoculars

The Wiley Science Editions

The Search for Extraterrestrial Intelligence, by Thomas R. McDonough

Seven Ideas That Shook the Universe, by Bryan D. Anderson and Nathan Spielberg

Space: The Next Twenty-Five Years, by Thomas R. McDonough

The Body in Time, by Kenneth Jon Rose

Clouds in a Glass of Beer, by Craig Bohren

The Complete Book of Holograms, by Joseph Kasper and Steven Feller

The Scientific Companion, by Cesare Emiliani

Starsailing, by Louis Friedman

Mirror Matter, by Robert Forward and Joel Davis

Gravity's Lens, by Nathan Cohen

The Beauty of Light, by Ben Bova

Cognizers: Neural Networks and Machines That Think, by Colin Johnson and Chappell Brown

Inventing Reality: Physics as Language, by Bruce Gregory

Planets Beyond: Discovering the Outer Solar System, by Mark Littmann

The Starry Room, by Fred Schaaf

Ozone Crisis: The 15-Year Evolution of a Sudden Global Emergency, by Sharon Roan

The Endangered Kingdom, by Roger DiSilvestro

Serendipity: Accidental Discoveries in Science, by Royston M. Roberts

Senses and Sensibilities, by Jillyn Smith

The Atomic Scientists: A Biographical History, by Henry Boorse, Lloyd Motz, and Jefferson Weaver

The Starflight Handbook: A Pioneer's Guide to Interstellar Travel, by Eugene Mallove and Gregory Matloff

Levitating Trains & Kamikaze Genes: Technological Literacy for the 1990s, by Richard P. Brennan

Journey Through Genius: The Great Theorems of Mathematics, by William Dunham

Reality's Mirror: Exploring the Mathematics of Symmetry, by Bryan Bunch

Touring the Universe through Binoculars

A Complete Astronomer's Guidebook

Philip S. Harrington

Wiley Science Editions

John Wiley & Sons, Inc.

New York • Chichester • Brisbane • Toronto • Singapore

To Dorothy, Wendy, and Helen:
Three generations of love and encouragement

Library of Congress Cataloging-in-Publication Data

Harrington, Philip S.
    Touring the universe through binoculars / Philip S. Harrington.
       p.   cm. — (Wiley Science editions)
    Includes bibliographical references.
    ISBN 0-471-51337-7
    1. Astronomy—Observers' manuals. 2. Binoculars. I. Title.
  II. Series.
  QB64.H37   1990
  523—dc20   90-35740

Printed in the United States of America

90 91 10 9 8 7 6 5 4 3 2 1

Preface

Star-gazing was never more popular than it is now. In every civilized country many excellent telescopes are owned and used, often to very good purpose, by persons who are not practical astronomers, but who wish to see for themselves the marvels of the sky. . . . And with the aid of an opera-glass most interesting, gratifying and, in some instances, scientifically valuable observations may be made of the heavens.

So wrote Garrett P. Serviss in the opening of his classic 1888 work, *Astronomy with an Opera-Glass*, the first book written about observing the universe through binoculars. It is fascinating how, a century after its publication, we may repeat his message with equal validity. With the recent launch of the Hubble Space Telescope, the promise of a rekindled U.S. space program, and an increased awareness of celestial events, public interest in astronomy has never been greater. Many people now realize that there is a lot going on over their heads.

Astronomy may be the oldest science, but it is anything but static. Today, our picture of the universe is so radically different from those earlier times, that Mr. Serviss and his contemporaries could not even begin to imagine it. A century ago, astronomers spoke of canals on Mars and the possibility of a "Planet X" beyond Neptune and another within the orbit of Mercury. They were perplexed by the strange spiral-shaped clouds that were seen across the sky. Many were found, but their origin and significance were not understood.

These riddles have been solved over the past hundred years. The Martian canals and the mysterious innermost planet are both dismissed as misinterpreted observations. The enigma of Planet X was answered in 1930 with the discovery of Pluto. Finally, with the application of photography to astronomy, those odd pinwheel nebulae were resolved into distant galaxies. Each was found to be a complete star system in itself, much like the Milky Way galaxy.

Today, other questions have replaced those of our forebears. We yearn to find out more about pulsars, quasars, and black holes. Did the universe originate from a colossal explosion 18 billion years ago? Will it continue to expand forever, or will it stop and reverse direction?

Like children, we have just started to take our first unsteady steps off of our own world and out into the Solar System. As space probes answer our questions about our celestial neighborhood, they raise others we never even thought to ask. Some will have to wait for future missions to answer them; the remainder will have to wait for future generations. There is still so much to learn.

Large telescopes and space probes are not needed to discover the heavens. Even the simplest binoculars will begin to reveal sights that were unsuspected or unrecognized in the days of Garrett Serviss.

Touring the Universe through Binoculars is intended to be the most thorough examination of the binocular sky ever compiled. Throughout its pages, you will find a wealth of objects and projects for amateurs who, like me, have made a conscious choice to use binoculars. To tell you the truth, I really wrote this book for myself! There is an abundance (dare I say, overabundance) of books on the market today that sing the praises of large-aperture, sophisticated instruments. Most of these books are quite good, but their intent is not for binocular observers.

Beginning with the Moon, we tour the Solar System and then escape the realm of our Sun to visit the stars. Over 1,100 deep sky objects visible through binoculars are listed. Of these, the appearance of more than 400 have been described in detail. Now, I am not saying that all are seen better through binoculars than through a telescope; that simply is not the case. Still, many are best viewed using low power and a wide field of view, and some cannot even be seen at all through the restrictive eyes of telescopes.

I welcome your comments, especially if you should find any errors (Gadzooks)! Just write to me in care of the publisher, John Wiley & Sons, Inc. I shall endeavor to answer all letters, but in case I miss yours, thank you in advance!

Gaze skyward on the next clear evening. The universe awaits.

Acknowledgments

While the book that you hold before you is the product of one pen, I wish to take a moment to thank the many individuals who have contributed to its ultimate success. Throughout the book, you will find what I judge to be some magnificent astrophotographs. These marvelous pictures were taken by Jim Barclay, Lee Coombs, Dennis diCicco, James Fakatselis, Martin Germano, Johnny Horne, Jeffrey Jones, Brian Kennedy, Jack Newton, George Viscome, Bernard Volz, and Kim Zussman. Take it from me, they are among the best amateur astrophotographers around today, and I thank them for their input.

I also wish to thank Jerry Burns, John Riggs, Pearson Menoher, Norman Butler, and Lee Cain. They constructed the unique binoculars and binocular mounting systems which you will find detailed in Appendix B.

I wish to extend my sincere appreciation to my proofreaders Richard Sanderson, Jack Megas, and Edward Pascuzzi, who read the manuscript over and over again and so eloquently passed on suggestions while dealing with my fragile ego. Aiding me in many different ways "behind the scenes" were Eric Hilton, John Kamon, David Eicher, Frederick Bump, and Louis Renzulli. Many thanks also to David Sobel and Frank Grazioli of John Wiley & Sons, and to Laura Cleveland of WordCrafters, for their patient guidance and help to this virgin book author.

Finally, to my wife Wendy, who proofed the final manuscript and provided endless encouragement over the years it took to compile the book: I simply could not have done it without you!

Contents

1   Why Binoculars?   1

2   The Moon   3

3   The Planets   25

4   Minor Members of the Solar System   37

5   The Sun   47

6   Stellar Happenings   55

7   A Survey of the Night Sky   81

Appendices

A   Caveat Emptor! (Let the Buyer Beware!) 263

B   Mounting Concerns 269

C   Care, Maintenance, and Other Tidbits 276

D   Binocular Manufacturers 279

E   Converting Universal Time to Local Time 281

F   For Further Information . . . 283

G   Bibliography 285

Index   287

1

Why Binoculars?

Why bother with binoculars? Today's amateur astronomers have a vast array of telescopic equipment and accessories from which to choose. Huge telescopes, advanced optical designs, and special accessories that were once considered to be only in the realm of the professional are now readily available to the hobbyist. With all this, why would anyone want to use plain old binoculars?

The answer is that, for all the diversity of instrumentation, one of the most useful, yet often neglected instruments to tour the universe with is a pair of binoculars. Their low power and wide field of view make binoculars ideal for either a casual scan or some pretty sophisticated observing.

Research has shown that when it comes to viewing the heavens, two eyes are definitely better than one. Our power of resolution and ability to detect faint objects are dramatically improved by using both eyes. In addition, color perception and contrast are enhanced.

Why take my word for it? On the next clear, moonless night, try an easy test. With both eyes open, cover one with your hand and look at the sky. Make a mental note of the faintest stars you see. Now uncover your eye and look again at the same area of the sky. Lo and behold, there are more stars! Indeed, it is not unusual to experience a 10-percent improvement in perception when viewing with two eyes instead of one.

Now repeat this test with a nebulous object, such as the Milky Way's hazy band stretching across the starry vault. Alternately cover and uncover one eye as before. The contrast between the soft glow of our galaxy's star clouds and the background sky will appear far more distinct through both eyes than with only one. In fact, many observers enjoy up to a 40-percent increase in the contrast of hazy objects merely by using both eyes. Note that while these two tests have been conducted without optical aid, similar results are achieved through binoculars versus a single equivalent-size monocular or telescope.

How can this improvement be explained? Light entering the eyes is focused by the lens onto the retina. The retina, consisting of rods and cones, converts the image into electrical pulses and sends them to the brain. The brain interprets the pulses and produces the image we see. By having the brain rely on only one set of pulses (i.e., by using just one eye), inconsistencies in the signals will interfere with the final image seen. With two sets of signals to interpret, the brain is able to reduce interference by averaging the pair of electrical messages. The result is the ability to see fainter, lower contrast objects.

There is little doubt that our view of the night sky improves just by using both of our eyes. This, along with their portability, affordability, and ease of operation makes binoculars the instrument of choice to tour the universe with.

2

The Moon

Lying an average of 238,000 miles away, the Moon offers more detail to earthbound observers than any other member of the Solar System. Towering mountains, large plains, abysmally deep valleys, and scores of craters are all easily visible in binoculars.

Most of us can still remember the anxious anticipation we felt just before our first view of the Moon's surface through either binoculars or a telescope. Its harsh beauty is awesome. Yet, powerful telescopes are not required to see many different lunar features.

Some lunar phases are better for sightseeing than others. Little or no detail may be spotted for the first couple of nights after the New Moon, but as the Moon progresses night after night toward the east, surface features become plainly visible. The later waxing-crescent through First-Quarter phases display a tremendous variety of lunar terrain for observers to marvel at. Dominating the equatorial zone are the vast expanses of the lunar seas Mare Crisium, Fecunditatis, Tranquillitatis, and Serenitatis. To their north are many scattered large craters, while to their south lies the Moon's "no man's land," that is, very rugged terrain (many mountains, craters, etc.—no manned landings due to ruggedness). The south polar region is awe inspiring in its coarse beauty, with craters so numerous that it is often difficult to distinguish one from another.

It is always fascinating to watch the unusual lighting effects along the Moon's terminator as the Sun rises and sets across the stark

Figure 2.1
The one-day-old Moon, as captured on April 6, 1989, by Jeffrey Jones.
He employed a 200-mm f/4.5 telephoto lens and a one-second exposure
on hypered Technical Pan 2415 film.

lunar surface. Depending on the angle, sunlight may just strike a
crater's rim, causing the crater to look like a bright, bottomless ring.
As the Sun's elevation increases, light travels down the steep, clifflike
walls until it floods the crater's floor.

As the Sun sinks in the lunar sky, mountain ranges seem to
experience metaphysical changes. First, the setting Sun thrusts their
bases into darkness. Then, ever so slowly, as the Sun continues to
approach the horizon, its diminishing light moves up the mountain-
side. Just before disappearing entirely, sunlight strikes only the highest
peaks, causing them to "detach" from the Moon and float above the
dark leading edge of the terminator.

In the past few years, an informal sport of spotting the extremely
thin crescent phases (Figure 2.1) has become popular among amateur
astronomers. Binoculars prove invaluable in this quest, since the very
young or very old Moon lies extremely low in the bright twilit sky.

The best months for spotting a very young Moon are April and
May in the northern hemisphere, and October and November in the

southern hemisphere. If you want to find the Moon less than a day before New Moon, then your best chance from north of the equator exists during July and August, while January and February are favored from points south of the equator. It is at these times that the ecliptic is nearly perpendicular to the horizon, placing the Moon higher in the sky. Slowly scan above the Sun's point below the horizon. If your view is free of trees, clouds, and other obstacles, you just might catch a glimpse of the very thin crescent.

This guide to the Moon presents a wide selection of lunar features that are resolvable through *steadily supported* (i.e., tripod-mounted) binoculars. The lunar maria, craters, mountain ranges, and miscellaneous features are listed in alphabetical order. The list is divided into two groups on the basis of location to the east or west of the Moon's central meridian (the imaginary line passing from the lunar north pole, through the middle of the visible disk, to the south pole). Those features found to the lunar east of the central meridian are best seen between New Moon and First Quarter (Figure 2.2), while those found to the lunar west of the central meridian are best seen between Last Quarter and New Moon (Figure 2.3). The quarter phases were chosen as cutoff points because they offer the best compromise between shadow relief and the amount of surface illuminated, while avoiding the overwhelming brightness of the larger phases.

In an effort to keep the finder charts as uncluttered as possible, the features are keyed to numbers or letters, which are also found in brackets at the beginning of each highlighted description. Note that lunar maria are specified by capital letters (A, B, etc.), craters are listed by number (1, 2, etc.), and mountain ranges and other features are denoted by lowercase letters (a, b, etc.).

To help the reader further to view specific lunar features at their prime, two numbers are listed in parentheses after each entry. These numbers signify the nights after New Moon that the particular target is most favorably placed for observation. In most cases, they indicate the time when the lunar feature is on or near the terminator, with the first number referring to sunrise and the second indicating sunset. All of the features thus referenced are actually visible for several nights during each lunar cycle.

Due to the Moon's brightness, regardless of phase, your eyes will never become fully dilated when observing the Moon. Therefore, binoculars with smaller exit pupils—as small as 2.5 mm—are perfectly acceptable for lunar study. Indeed, the large 7-mm exit pupils of night glasses transmit so much light that the observer's eyes will be unnecessarily taxed. Consult Appendix A for a discussion of calculation of exit pupils.

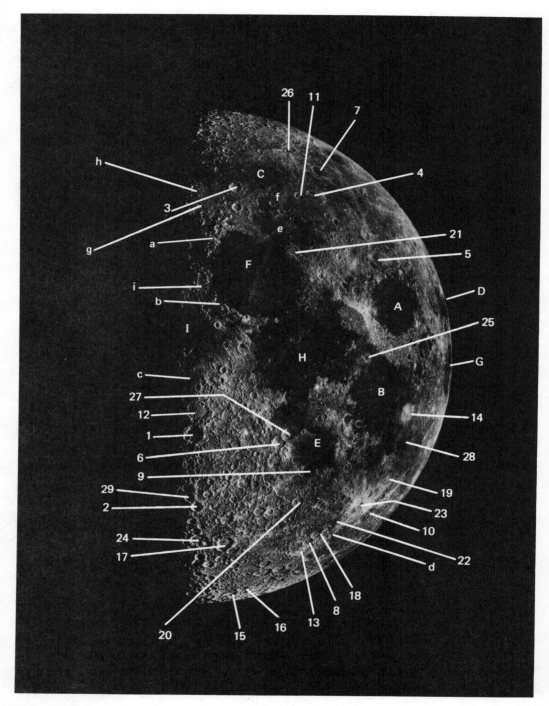

Figure 2.2
Map of the First-Quarter Moon showing the locations of features highlighted in the section of the text titled "New Moon through First Quarter." Lick Observatory photograph.

6

New Moon Through First Quarter

Maria

Mare Crisium [A], the Sea of Crises, is a large oval plain measuring 270 miles by 350 miles, with the long dimension running east to west. This is just the opposite of the visual impression we get from Earth because of the foreshortening of the lunar globe. Unlike the other, interconnecting maria, Mare Crisium stands alone. While no manned Apollo mission landed there, the Sea of Crises was visited by three unmanned Soviet spacecraft. Luna 15 was the first to land, on July 16, 1969 (only three days before Apollo 11 landed in Mare Tranquillitatis), followed by Luna 23 five years later. The third spacecraft, Luna 24, landed in 1976, drilled 18 inches below the surface, and returned a cylinder of lunar soil to Earth. This mission holds the distinction of being the last in which a spacecraft visited the Moon.

Mare Fecunditatis [B], the Sea of Fertility, is seen to the south of Mare Crisium. Only the twin craters Messier and Messier A blemish its smooth, dark surface. Little attention was paid to the area during the days of the lunar probes. Only Luna 16 nestled down on this barren plateau back in 1970. It became the first unmanned Soviet mission to bring soil samples back to Earth.

Mare Frigoris [C], the Sea of Cold, is most unusual in appearance. Instead of being the typical circular plain, Mare Frigoris stretches for over 700 miles, but is no more than 45 miles wide at points. While no large craters lie within, several striking examples—including Endymion, Atlas, Hercules, Aristoteles, and Plato—surround it.

Mare Marginis [D], as the name implies, is only marginally visible. This oddly shaped mesa measures 210 miles by 290 miles. Most earthbound observers are unfamiliar with Mare Marginis, as only a small portion of it is visible along the eastern limb, and then only during favorable librations. Look for it directly east of Mare Crisium.

Mare Nectaris [E], the Sea of Nectar, spans a 220-mile by 265-mile area, making it one of the smallest of the Moon's major maria. Found off the southern shore of Mare Tranquillitatis, it is encircled by several prominent craters, notably Theophilus and Fracastorius.

Mare Serenitatis [F], the Sea of Serenity, is thought by some authorities to be the oldest lunar mare of all. Covering 360 miles by 420 miles, it is bound by the Haemus Mountains to the south, the Apennines to the west, the Caucasus Mountains to the north, and the Taurus Mountains to the east. No mission from Earth has ever touched its nearly crater-free surface.

Mare Smythii [G] hides along the Moon's eastern limb and is almost always a difficult catch. At best, it may be seen as a long, thin patch of darkness south of Mare Crisium and Mare Marginis. Though impossible to tell from our vantage point on Earth, Mare Smythii covers 165 miles east to west by 255 miles north to south.

Mare Tranquillitatis [H], 400 miles by 550 miles in extent, became the most famous lunar sea after Apollo 11 landed there on July 19, 1969. As we look toward the Sea of Tranquility, it is easy to recall Neil Armstrong's words: "Houston, Tranquility Base here. The Eagle has landed." Later, as he became the first person to set foot on another body in space by taking "one small step for [a] man . . . one giant leap for mankind," Mare Tranquillitatis became ingrained in history and astronomy books alike.

Armstrong and lunar module pilot Edwin "Buzz" Aldrin set their craft down in the northwest corner of the Sea, to the south of the prominent crater Theophilus. Other spacecraft that have visited the Sea of Tranquility include Ranger 6, Ranger 8, and Surveyor 5.

Mare Vaporum [I], the Sea of Vapors, is a small plain wedged between the Haemus and Apennine Mountains. Only the crater Manilius is large enough to mar its surface through most binoculars. The most powerful glasses should also spot the Hyginus Rill crossing the southern portion of the sea. It appears as little more than a thin pencil line, but in reality measures 140 miles long by 2 miles wide.

Craters

Albategnius [1], measuring 81 miles in diameter, is an outstanding crater through binoculars. A noticeable off-center peak rises from the dark floor, some 14,000 feet below the crater's rim. (7,22)

Aliacenis [2] rides the terminator on the night of the First Quarter Moon. Paired with Werner, it stands out as a deep lunar "pothole" spanning 46 miles by 55 miles. Its sharp crater walls plummet for nearly 2½ miles to a relatively dark, smooth floor. (7,21)

Aristoteles [3] is a prominent 55-mile-diameter crater found on the southern edge of Mare Frigorus. Its steeply banked walls plummet 12,000 feet to a smooth floor. (6,20)

Atlas [4], at 54 miles across, is one of the more noteworthy craters visible during the waxing crescent phases. A multiple peak ascends 2,600 feet above its 10,000-foot-deep floor. Atlas forms a nice pair with Hercules, a slightly smaller crater to its west. (5,18)

Cleomedes [5] is one of the most visible craters on the waxing crescent. This out-of-round crater, located just to the north of Mare Crisium, measures 81 miles by 92 miles. Its sheer walls descend nearly three miles to a rough floor. (3,16)

Cyrillus [6] is an extremely old crater found to the west of Mare Nectaris. Were it isolated, it would dominate the field, since it is 62 miles across and 11,800 feet deep. However, Cyrillus is overshadowed by Theophilus, a fresher crater that protrudes into its northeastern wall. (5,19)

Endymion [7] is a large, distinguished crater that is most easily seen during the middle of the waxing crescent phases. Found in the northeast quadrant between Mare Frigoris and the lunar limb, it is set apart from the immediate surroundings by its dark floor. Endymion is 77 miles wide, with walls that cascade for 16,100 feet. (4,18)

Fabricius [8] lies in the Moon's southeast district. It is found squeezed between Metius and Janssen, two larger enclosures. Fabricius must be the most recent of the three, as it juts out into the others. (4,18)

Fracastorius [9], though once a 73-mile-wide crater, appears as a bay on the south shore of Mare Nectaris. Its north wall was washed away when a flood of hot lava from the neighboring maria filled its floor. Through low-power binoculars, it appears as a complete ring, but the greater resolving power of telescopes reveals this to be an illusion caused by a series of hills and mounds. As it remains today, Fracastorius is one of the Moon's finest examples of a partially destroyed crater ring. (6,20)

Furnerius [10] is noted for its sharply angled rim, which is almost hexagonal in form. Measuring 81 miles across, this crater is one of the key features visible on the very young waxing crescent, yet is all but invisible once the ray system of neighboring Stevinus catches the Sun. (2,16)

Hercules [11] and previously mentioned Atlas form a powerful pair of craters near Mare Frigoris. The smaller of the two, Hercules is 45 miles across with pronounced walls and a distinctive dark floor. (5,18)

Hipparchus [12] is found just north of Albategnius. While Hipparchus is the larger of the two craters, at 83 miles by 89 miles, it proves less noteworthy. The years have been hard on its ancient face, with the 7,500-foot-deep walls bearing the scars of more recent impacts. (7,21)

Janssen [13] is a large, very old crater bordering Fabricius to the south. Though Janssen measures 122 miles by 153 miles, its battered walls mask the crater's identity, except under ideal lighting conditions. (4,24)

Langrenus [14], found on the southeastern shore of Mare Fecunditatis, is an amazing crater to watch from waxing crescent to Full Moon. As the Sun rises higher in the sky above it, 85-mile-wide Langrenus seems almost to catch fire as its floor is transformed from a dull gray to a brilliant whitish glow. (2 through 16)

Manzinus [15] is found toward the Moon's southeast limb. Measuring 55 miles by 63 miles, it forms an attractive crater pair with **Mutus [16]** to the east. (5,19)

Maurolycus [17] spans 73 miles and has walls 16,700 feet high. It is found in the rugged highlands toward the Moon's southern limb, where it dominates the view. (7,14)

Metius [18] lies in the southeastern region of the Moon, directly adjacent to Fabricius. Together, the two craters, measuring 53 and 48 miles across, respectively, look like twin depressions through low-power binoculars. Careful scrutiny through higher power glasses reveals a strong central peak in Fabricius, while the floor of Metius appears comparatively smooth. (4,18)

Petavius [19], a 99-mile by 110-mile enclosure, is one of the highlights of the young crescent Moon. Look for it to the south of Mare Fecunditatis. When the Sun is at just the right elevation, the 8,200-foot-high central peak of the crater is easily seen in 7× glasses. (2,16)

Piccolomini [20] sees sunrise about four days after New Moon and is an outstanding crater from then until past First Quarter. Measuring 54 miles across, it exhibits a strong central peak through 11× and higher power binoculars. Look for Piccolomini to the south of Fracastorius and Mare Nectaris. (5,18)

Posidonius [21] stands out well as a 52-mile by 61-mile chasm. Bordering Mare Serenitatis to the south, it was partially filled with lava during the Moon's volcanic era some three billion years ago. (5,20)

Rheita [22] is a small, easily overlooked crater in the southeastern lunar highlands. Though only 44 miles across, its presence helps point the way to the Rheita Valley (see "Other Features"). (4,18)

Stevinus [23] is hardly visible through binoculars when the Sun rises over it on the third day after New Moon. However, a day or two afterwards, the whole region catches fire as the ray system from an unresolvable crater (known as Stevinus A) becomes sunlit. Forty-eight-mile-wide Stevinus appears framed by these brilliant rays. (5 through 9)

Stofler [24] is a large oval crater that rides the terminator at the quarter phases. Measuring 68 miles by 85 miles, its southeastern wall was obliterated millennia ago by the smaller impact crater Faraday. (7,21)

Taruntius [25] blemishes the northwest corner of smooth-surfaced Mare Fecunditatis. Though only 36 miles in diameter, its bright appearance contrasts nicely with the dark surrounding plain. (4,18)

Thales [26], when viewed at or near Full Moon, forms one-half of a pair of "spotlights" on the northern limb of the Moon. The other beacon is Anaxagoras, described in the section titled "Last Quarter through New Moon." Thales, at 22 miles across, is barely discernible at phases other than Full Moon. (13 through 17)

Theophilus [27] is found on the northwest shore of Mare Nectaris and dominates the area. The strong, jagged walls of 65-mile-wide Theophilus cascade down for over four miles to a rough floor. (5,19)

Vendelinus [28] is an old crater found to the south of much more prominent Langrenus and covers 92 miles by 100 miles. Its northeast wall has been demolished by the smaller impact crater Lame. (2,16)

Werner [29] teams with Aliacensis to form an impressive crater duet on the night of the First Quarter. While both stand out well when on the terminator, they are easily lost within a couple of nights as the Sun rises higher in their sky. (7,21)

Mountain Ranges

The **Caucasus Mountains** [a] are a conspicuous range that divides Mare Imbrium from Mare Serenitatis. With their highest peaks rising 17,400 feet above "sea level," these mountains are an especially impressive sight when found on the terminator. (6,20)

The **Haemus Mountains** [b], a little-known range of peaks that rise to about 10,000 feet, form the southern border of Mare Serenitatis. They appear to merge with the Apennines to create a peninsula that separates the Sea of Serenity from the Sea of Tranquility. (6,20)

Other Features

Sinus Medii [c], a large oval plain measuring 100 miles by 200 miles, is found at the center of the Moon's near side. The landing site for both Surveyor 4 and Surveyor 6 in the mid-1960s, its smooth surface is unscarred by large craters. (7,21)

The **Rheita Valley** [d] is a puzzling feature that stretches over 250 miles to the southeast from the crater Rheita. The valley appears as a giant scar on the Moon's surface and stands out prominently when sunlight has not quite reached its bottom. (4,18)

Lacus Somniorum [e], the Lake of Dreams, appears as an oddly shaped bay extending northward from Mare Serenitatis. Though darker than the enclosing highland area, its surface is noticeably lighter than the maria. Farther north, Lacus Somniorum floods into **Lacus Mortis** [f], the Lake of the Dead, which in turn passes into Mare Frigorus. Lacus Somniorum measures about 90 miles by 180 miles, while Lacus Mortis spans 100 miles by 110 miles. Both are littered with many "islands." (4,18)

Last Quarter Through New Moon

Maria

Mare Cognitum [J] is a small flat plain measuring only 120 miles by 200 miles and found to the south of the crater Copernicus. It is separated from the Ocean of Storms to the west by the Riphaeus

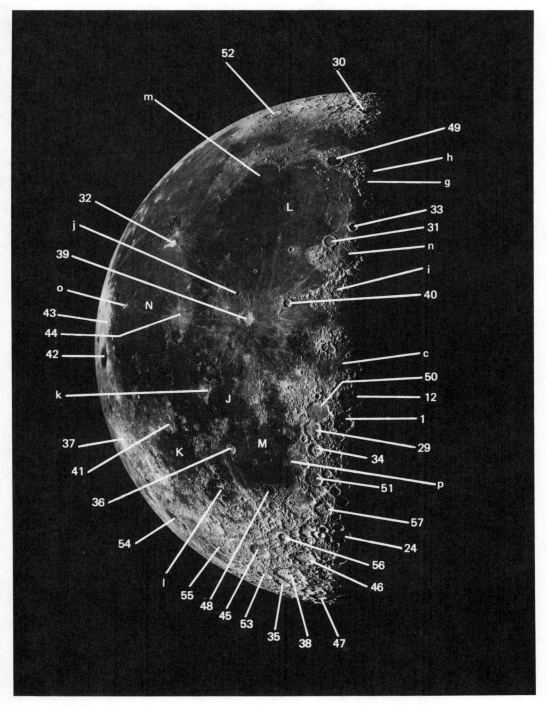

Figure 2.3
Map of the Last-Quarter Moon showing the locations of features highlighted in the section of the text titled "Last Quarter through New Moon." Lick Observatory photograph.

Mountains and bordered on the east by a small "island" of craters. Though difficult to make out through binoculars, the ancient crater Fra Mauro forms the northern edge of this island. Fra Mauro was the site of 1971's Apollo 14 mission.

Mare Humorum [K], the Sea of Moisture, is a nearly circular dark plain to the south of Oceanus Procellarum. Its 250-mile by 275-mile surface appears quite smooth through binoculars, except for a few "waves" to the east of center. Some experts believe that Mare Humorum is the oldest of the maria.

Mare Imbrium [L], the Sea of Rains, spans 670 miles by 750 miles. Touched by only the Luna 17 mission, it is highlighted by many striking features, such as Sinus Iridum, the craters Archimedes and Eratosthenes, and the surrounding Apennine and Alps Mountains.

Mare Nubium [M], the Sea of Clouds, is notable for its unusually dark floor. Some minor blemishes may be seen through binoculars, but none is large enough to be easily identifiable.

Oceanus Procellarum [N], the Ocean of Storms, is the largest of the lunar plains and dominates the waning phases. It encompasses more than one million square miles of the Moon's surface. Several noteworthy features, including the craters Aristarchus, Kepler, and Copernicus, are found within its vast borders. Among the many spacecraft to land there were Surveyor 1, Surveyor 3, and Apollo 12 from the United States, as well as the Soviets' Lunas 5, 7, 8, 9, and 13.

Craters

Alphonsus [29] is one of five large craters that stretch southward from Sinus Medii at the Moon's center. Alphonsus spans 64 miles by 73 miles. The site of the Ranger 9 impact mission in 1965, the crater has been observed to have, on rare occasions, a cloud of carbon molecules temporarily floating above it. Photographs of this unusual phenomenon taken in the 1950s prove that the Moon is not a dead world. (8,22)

Anaxagoras [30], when on the terminator, is far surpassed by many of its larger neighboring craters. Yet, by the time the Sun has risen high in its sky, this 33-mile-wide crater puts on a striking display of bright rays that gives it prominence. Together, Anaxagoras and Thales, described earlier, form an impressive pair of "spotlights" along the Moon's northern limb. (13,17)

Archimedes [31] is found in eastern Mare Imbrium and spans 51 miles. A portion of its walls were flooded by lava during the Moon's volcanic era. (8,22)

Aristarchus [32] is one of the smallest craters identifiable through binoculars. Surprisingly, it is also one of the easiest to identify, as a magnificent system of rays sprays out from its rim into surrounding Oceanus Procellarum. Indeed, these rays are so dazzlingly bright that they can completely overwhelm the crater itself. (11,25)

Aristillus [33], found to the southeast of the Alps in Mare Imbrium, is 36 miles across. As the Sun rises higher in the lunar sky, a system of bright rays appears to spray from the crater into the surrounding plains, making Aristillus easy to find. (8,21)

Arzachel [34] is found just south of Alphonsus and is a second member of the prominent north-south collection of craters extending southward from Sinus Medii. It measures 59 miles wide, with sheer walls that fall 13,000 feet to a jagged floor. (8,22)

Blancanus [35], to the south of mighty Clavius, is a smooth-floored crater spanning 63 miles by 66 miles. (9,23)

Bullialdus [36] is a small, but prominent, crater located on Mare Nubium. Thirty-eight miles in diameter, it stands out well against the surrounding dark mare floor. Pay close attention to its strong, bright central peak and terraced walls, which cause petite Bullialdus to resemble the powerful crater Copernicus. (9,23)

Byrgius A [37] is the smallest feature of those listed. Spanning only 10 miles by 13 miles, it would normally be passed right over without a second glance were it not for its magnificent system of brilliant rays. These rays, easily seen from Full Moon to the Waxing Crescent, have been traced for over 260 miles. They completely overwhelm the larger crater Byrgius, which lies to the smaller crater's southwest. (14 through 23)

Clavius [38], second largest crater on the Moon's earthbound side, is one of the easiest to recognize. Its huge 132-mile by 152-mile walls are especially prominent right after the quarter phases. Then, sunlight will just catch the top of the rim, causing a bright ring to protrude into the cold, dark lunar night. On the floor of Clavius are several smaller impact craters, with 30-mile-wide **Clavius B** the most obvious. (8,22)

Copernicus [39], spanning 60 miles, is another easily found crater. Located where Mare Imbrium meets Oceanus Procellarum, Copernicus has a brilliant ray system that explodes into view against the darker background of the maria. Its starburst pattern is unmistakable in even the most modest glasses. The central mountain peak of Copernicus may be glimpsed in 7× binoculars. (9,22)

Eratosthenes [40] is readily encountered halfway between Copernicus and the Apennine Mountains. The steeply banked walls of this 37-mile-wide crater drop nearly 2½ miles to a rough floor. (8,22)

Gassendi [41] is a large, rather shallow crater perched on the northern edge of Mare Humorum. With a rim spanning 70 miles, this crater's 3,600-foot-high central mountain stands out in 10× and larger binoculars when the lighting is right. (11,25)

Grimaldi [42] is to the waning crescent what Mare Crisium is to the waxing crescent. Found on the extreme western limb, Grimaldi appears as a decidedly oval dark patch separated just a bit from Oceanus Procellarum. This elliptical shape is only an illusion caused by the Moon's curvature, as Grimaldi is nearly circular at 140 miles by 145 miles across. (13,27)

Hevelius [43], a 66-mile by 69-mile enclosure just to the north of Grimaldi, is another highlight of the "Old Moon." Intruding into its north wall is a smaller crater named **Cavalerius**. At 37 miles across, Cavalerius will prove challenging in most glasses. (13,27)

Kepler [44], though only 20 miles across, is one of the most prominent craters on the Moon. Situated near the center of Oceanus Procellarum, it has a bright ray pattern and will consequently remind many of a miniature Copernicus, which, incidentally, is just to the east. Because they are completely isolated from the bright highlands regions, the rays from both Kepler and Copernicus may even be seen with the unaided eye. (11,25)

Longomontanus [45] is seen to the northwest of Clavius and southwest of Tycho, while another obvious crater, **Maginus [46]**, is found an equal distance east of Clavius and Tycho. Collectively, all four create a prominent diamond pattern of craters in the Moon's southern quadrant. Longomontanus, at 107 miles across, and Maginus, measuring 100 miles by 115 miles, are two of the largest craters found on the Moon's near side. (8,22)

Moretus [47] is found between Clavius and the Moon's southern limb. This area is one of the most rugged on the lunar face, which makes individual craters difficult to identify. At 75 miles across, Moretus is one of the largest craters found in the region, yet it will still present a challenge to most observers. (8,22)

Pitatus [48] protrudes into the southern rim of Mare Nubium. Its floor is darker than those of most craters, although not as dark as the maria themselves. Under the right lighting conditions, the bright 60-mile by 69-mile walls of Pitatus may be seen to encircle its shaded floor like a silver ring. (8,22)

Plato [49] is a striking feature on the Last Quarter Moon. Its floor is the darkest of any crater found on the Moon's Earth-facing side and appears even darker than neighboring Mare Imbrium and Mare Frigorus. Set in the foothills of the Alps, Plato looks far more oval than its 64-mile by 67-mile dimensions imply. This effect is due to the visual foreshortening of the lunar globe. Combining its stark floor with the attractive ruggedness of its surroundings, Plato highlights one of the most enticing lunar regions. (9,22)

Ptolemy [50] is the largest in the line of craters that includes Alphonsus, Arzachel, Purbach, and Walter. Ptolemy is 93 miles across, with walls that fall nearly 10,000 feet to a smooth surface. The central peak seen through telescopes cannot be glimpsed through binoculars. (8,22)

Purbach [51] belongs to the line of five craters set near the terminator on the night of Last Quarter. Purbach is slightly oval, at 62 miles by 73 miles across, and is the most difficult of the five to resolve. (8,22)

Pythagoras [52] is a prominent crater set between Mare Frigoris and the northwestern limb. Covering 81 miles by 90 miles, the multiple terracing effect of Pythagoras' walls may be spotted in high-power glasses, along with the faintest hint of a central mountain. (12,26)

Scheiner [53] forms a triangle with Clavius and Blancanus. With a rim 71 miles in diameter, it exhibits a smooth floor through binoculars. (9,22)

Schickard [54] is a unique crater spanning 135 miles by 150 miles and found near the Moon's southwestern limb. It has an oddly shaded floor, with two darker regions sandwiching a higher and brighter central plateau. Selenologists are not sure exactly what caused this apparent sinking of part of Schickard, but the effect is striking under the right lighting conditions. (12,26)

Schiller [55] is a long, footprint-shaped enclosure 48 miles wide by 113 miles long. Seen near the southwestern limb, its elongation is even more pronounced due to the curvature of the Moon's globe. To this day, the origin of Schiller remains a mystery. The only thing selenologists agree on is that it almost certainly did not result from a single impact. (11,25)

Tycho [56], though only 56 miles across, transforms into the most spectacular crater of all during the gibbous and full phases. Dazzlingly bright rays scatter radially outward from the crater for hundreds of miles, overwhelming all other features on the Moon. Perhaps the youngest of the Moon's major craters, Tycho was the landing site of Surveyor 9 in 1968. The last of the pre-Apollo series, Surveyor 9 transmitted thousands of pictures back to Earth during the two-week span of its mission. (13,24)

Walter [57] is the smallest and southernmost member in the line of five craters stretching southward from Sinus Medii. An oval depression spanning 78 miles by 88 miles, Walter possesses an unusual off-center mountain found in the northeast part of the crater floor. Under the right lighting, the peak's shadow should provide enough relief for detection in 10× glasses. (8,22)

Mountain Ranges

The **Alps** [g], like their earthly namesake, are an impressive mountain range. They stretch for 350 miles, from Mare Frigorus in the north to the shores of Mare Imbrium. The highest peak in the lunar Alps has a measured altitude of 14,000 feet. During the waning gibbous phases, the striking crater Plato will help point your way to these slopes. Of special interest is the **Alpine Valley** [h], a gap that cuts straight through the mountains, acting as a natural canal connecting the two lunar seas. (8,21)

Bordering the southeastern shore of Mare Imbrium are the **Apennine Mountains** [i]. Many outstanding peaks may be seen dotting this range's 450-mile length, including Mount Hadley, site of the Apollo 15 landing in 1971. (8,22)

The **Carpathians** [j] make up a 200-mile-long mountain range. They are easily found just north of the impressive crater Copernicus and contain peaks rising 6,600 feet above lunar sea level. (9,23)

The **Riphaeus Mountains** [k] act as a natural border between Mare Cognitum and Oceanus Procellarum. Although resolution of individual peaks is an arduous task, they collectively appear as a bright arc some 30 miles wide and 110 miles long. (10,24)

Other Features

Palus Epidemiarum [l] is a large plain extending to the southwest of Mare Nubium. It is separated from the Sea of Clouds by one of Tycho's bright rays. Measuring 110 miles by 175 miles, it is pockmarked with many small craters. Forty-seven-mile-diameter **Capuanus** is the largest of these infiltrators. (10,24)

Sinus Iridum [m], the Bay of Rainbows, is a favorite feature of the Moon. Originally, Sinus Iridum was a complete crater, but its southern wall was totally washed away when lava from Mare Imbrium crashed against it. Two promontories, **Heraclides** and **Laplace,** mark the opening of the bay's 160-mile-wide mouth, while the **Juras Mountains** form its northern perimeter. (10,24)

Palus Putredinus [n] is a small, uneven plain found close to the crater Archimedes. Just outside the border shared with the Apennine Mountains, near Mount Hadley, is the site of 1971's Apollo 15 mission. (8,21)

Reiner Gamma [o] is a bright marking between Hevelius and Kepler. It is not thought to be associated with any visible surface feature. Never casting a shadow, Reiner Gamma apparently lies on the same level as the surrounding plain. It measures 27 miles by 50 miles and actually has a measurable magnetic field. Reiner Gamma is a most unusual feature, indeed! (13,26)

The **Straight Wall** [p] is one of the Moon's most fascinating surface features. Seemingly rising out of nowhere, this 75-mile-long cliff suddenly climbs at a 40° angle to a height of 1,200 feet. Although the Straight Wall itself will not be seen, its dark, pencil-thin shadow may be spotted on the Moon's eighth day. Check back in two weeks, when the Sun's light will fully illuminate the then-shadowless white cliff. (8,23)

Lunar Occultations

Over the course of every lunar cycle, the Moon will apparently pass in front of, or *occult,* many distant stars. Viewing these occultations can be an enjoyable and useful activity for amateurs to engage in. By timing their occurrences precisely, scientists may detect slight fluctuations in the Moon's orbital distance and speed. It is also possible during an occultation to determine whether the occulted star is a single or double star.

On comparatively rare occasions, the Moon will pass in front of planets and asteroids as well (see Figure 2.4). Watching the Moon occult a planet is especially interesting. Rather than blinking out suddenly like a star, the planet will be seen to slowly fade away over a period of a few seconds as the Moon's limb gradually blocks the planet's disk.

Predictions of occultations are computed and published annually by the U.S. Naval Observatory (USNO) in Washington, D.C. and are available for 18 standard "stations" found across the United States and Canada. Choose whichever station is closest to your observing site, and use the predicted time for that station as a guide for when

Figure 2.4
Occultation of Venus by the waning crescent Moon. This photograph was taken by the author at the prime focus of an 8-inch f/7 reflecting telescope. The original exposure was $1/15$ second with ISO 400 film.

the occultation will occur. USNO predictions for the occultations of brighter stars are also available in the annual Royal Astronomical Society of Canada's *Observer's Handbook,* as well as in both *Sky and Telescope* and *Astronomy* magazines.

For more information on occultation timing, interested readers should contact the International Occultation Timing Association (IOTA) at 6 N 106 White Oak Lane, St. Charles, Illinois 60174. This organization can provide you with instructions and predictions of occultations for your location. Membership in IOTA includes occultation predictions, a subscription to *Occultation Newsletter,* and other material.

Lunar Eclipses

I always look forward to a lunar eclipse (Figure 2.5). In fact, I first became interested in astronomy a quarter century ago when I observed a total lunar eclipse through an old (now long gone) pair of 7× binoculars. I can still recall the rush of anticipation I felt as the Earth's coppery red umbral shadow crept ever so slowly over the Full Moon's face. Deeper and deeper the Moon glided into the darkness, until it was fully blocked from direct sunlight. The deep crimson light that illuminated the lunar surface (the result of sunlight bending through Earth's atmosphere) appeared more vivid through the binoculars than with the unaided eye. It was not until an hour later that the Moon slowly began to sneak out of Earth's shadow. The eclipse then drew to a close, but a fascination with the universe had just been born.

Like snowflakes, no two lunar eclipses are exactly the same. Sometimes, the umbra casts a bright red-orange hue onto the lunar surface, while other eclipses appear coppery red or even brownish gray. A total eclipse's color is dependent upon the clarity of our planet's upper atmosphere. In general, a vivid eclipse will be seen when the air is free of pollutants, while a darker total phase will be witnessed when the atmosphere is littered with foreign particles, such as after a volcanic eruption. During the lunar eclipses of December 30, 1963, and December 30, 1982, the Moon nearly disappeared from view at midtotality! Ash suspended high above Earth's surface from the 1963 eruption of Mount Agung on Bali and the 1982 explosion of Mexico's El Chicón respectively caused these darker-than-usual eclipses.

To rate the coloring of the umbral shadow, the French astronomer A. Danjon came up with the following five-point luminosity scale:

Figure 2.5
The Moon entering the Earth's shadow on December 30, 1982. Note the stars Castor and Pollux appearing on the top. With ISO 400 film, exposure times increased from ¹/₆₀ second at f/16 to ten seconds at f/1.7 as the Moon slipped deeper into the umbra. Exposures were taken at five-minute intervals. Photograph by the author.

$L = 0$:   Very dark eclipse, with the Moon almost invisible at midtotality.

$L = 1$:   Dark eclipse, gray or brownish coloration, details distinguishable only with difficulty.

$L = 2$:   Deep red or rust-colored eclipse, with a very dark central area in the shadow and the outer edge of the umbra relatively bright.

$L = 3$:   Brick-red eclipse, usually with a bright or yellow rim to the shadow.

$L = 4$:   Very bright copper-red or orange eclipse, with a very bright bluish shadow rim.

It is also interesting to estimate the Moon's brightness at midtotality. Normally, the Full Moon shines at magnitude −12.7, but this was seen to drop by a factor of nearly two million, to 3rd magnitude during the December 1982 eclipse!

To assess its magnitude, the Moon must be compared to stars or planets of known brightness. But how can we compare the ½° diameter lunar disk to a point of light? For an accurate estimate, the Moon must appear equal in size to the object it is compared with. Fortunately, this is an easy trick through binoculars. Simply turn the binoculars around and look through the objective lenses. Everything in the field, including the Moon, will decrease in size by the same factor as the binoculars normally increase images by. The Moon's magnitude may then be directly estimated by comparison with other sky objects of known brightness.

Table 2.1 lists upcoming partial and total lunar eclipses through

TABLE 2.1 Upcoming Lunar Eclipses, 1990–2000

Date	Type of Eclipse	First Contact (U.T.)	Mideclipse (U.T.)	Last Contact (U.T.)
9 Feb 1990	Total	17:30	19:12	20:54
6 Aug 1990	Partial	12:43	14:11	15:39
21 Dec 1991	Partial	10:01	10:33	11:05
15 Jun 1992	Partial	03:27	04:57	06:27
9–10 Dec 1992	Total	22:00	23:44	01:28
4 Jun 1993	Total	11:12	13:01	14:50
29 Nov 1993	Total	04:40	06:25	08:10
25 May 1994	Partial	02:39	03:31	04:23
15 Apr 1995	Partial	11:42	12:18	12:54
3–4 Apr 1996	Total	22:22	00:10	01:58
27 Sep 1996	Total	01:13	02:54	04:35
24 Mar 1997	Partial	02:59	04:40	06:21
16 Sep 1997	Total	17:08	18:46	20:24
28 Sep 1999	Partial	10:22	11:33	12:44
21 Jan 2000	Total	03:03	04:44	06:25
16 Jul 2000	Total	11:58	13:56	15:54

the year 2000. To find out whether a particular lunar eclipse will be visible from your location, simply convert the listed Universal Time (U.T.) of the event to your local clock time (see the appendices for more information). If the eclipse occurs during your night, then you are in luck; otherwise, better luck next time!

3

The Planets

Over the millennia since human beings first gazed skyward, the wandering planets have been a source of endless fascination. First revered as gods, the planets have only recently begun to shed their cloaks of mystery to roving unmanned interplanetary probes. Each of the five naked eye planets—Mercury, Venus, Mars, Jupiter, and Saturn—has been visited and studied at close range. Uranus has also been scrutinized by the passing eye of a spacecraft, which most recently flew by Neptune. Now, only isolated Pluto remains unvisited.

Since their formation nearly five billion years ago, each of the nine planets has developed a unique set of characteristics. As we stop at each on our tour of the Solar System, some of their traits will reveal themselves through our binoculars. Unfortunately, many more will remain hidden from view: because of the planets' relatively small size and great distances, their features are difficult to observe through common low-power binoculars. Brilliant Venus appears largest of all from Earth, and yet its apparent diameter measures only $1/30$ as large as that of the Full Moon.

High-power glasses, such as 20×50, 20×70, or 30×80, are recommended for the best view. Keep in mind that the exit pupil diameter is not as important a consideration in choosing binoculars for studying the brighter planets as it is for observations involving faint deep sky objects.

Mercury

Of the five naked eye planets, Mercury is by far the most difficult to glimpse. This tiny world is forever trapped in an orbit that holds it an average of only 36 million miles from the fiery Sun. As a result, Mercury is never farther than 28° from the Sun in our sky, and it is normally seen only in very heavy twilight, up to about an hour after sunset or before sunrise.

Earth-based telescopes reveal that Mercury goes through phases similar to our own Moon's, but these same telescopes are unable to show any distinct surface features of the planet. It was not until 1974 that Mercury finally showed us its true self. That was the year Mariner 10 flew by the planet and sent back spectacular photographs of a rugged, arid, crater-strewn surface.

Binoculars prove invaluable when searching for this elusive planet. First, check predictions to see when it is supposed to be visible in the sky. The ideal time to search for Mercury is when the planet is at "greatest eastern elongation" or "greatest western elongation." Mercury (and Venus, as we shall see later) is said to be at *greatest eastern elongation* when it is at maximum distance east of the Sun, which occurs in western twilight after sunset. *Greatest western elongation* occurs when Mercury (or Venus) is at its farthest distance west of the Sun, making it visible in the east just before sunrise.

The best opportunity for observers in the northern hemisphere to see Mercury is when it reaches greatest eastern elongation in the spring or greatest western elongation in the fall, as shown in Figure 3.1. Slowly scan back and forth just above the horizon where Mercury is expected to be. The planet may just be glimpsed as a bright, flickering point of light immersed in deep morning or evening twilight. Sky conditions must be quite clear to attain success. You should attempt to spot Mercury only when there is little or no smoke, fog, or clouds hugging the horizon. Even the slightest interference might render the planet undetectable.

Mercury takes only 88 days to travel once around the Sun. As the speedy planet races around our star, it alternately passes in front of and behind the Sun as seen from Earth. These points in its orbit are called *inferior conjunction* and *superior conjunction,* respectively.

Although the tilt of Mercury's orbit usually takes it either above or below the solar disk at conjunction, occasionally it lines up with the Earth and the Sun. When this occurs at inferior conjunction, Mercury appears to *transit* the Sun. Transits of Mercury are relatively rare. The last took place in 1986, although it was not visible from North America. The next transit that will be seen from this part of the globe is set for May 7, 2003.

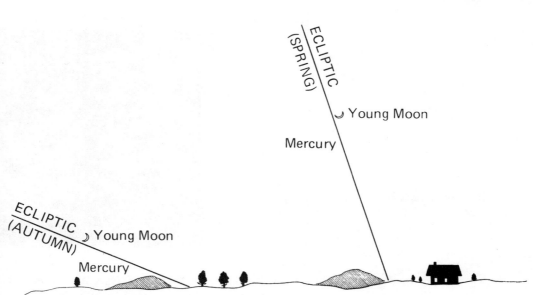

Figure 3.1
The angle of the ecliptic to the evening western horizon during the northern hemisphere's spring and autumn. Note that both the Moon and Mercury can appear significantly higher in the spring sky than in the autumn sky, even when they both lie the same angular distance away from the Sun. Diagram by the author.

Are binoculars sufficient to watch a transit? It is generally thought that at least 50× is required, but the most powerful binoculars, such as a pair of 30 × 80s, may be adequate to just make out tiny Mercury as it travels across the solar disk. Be sure to exercise precaution against the harmful rays from the Sun before looking at a transit. Chapter 5 gives information on proper safety measures for observing the Sun.

Venus

Venus, at times the closest planet to Earth, is dazzling when suspended high in morning or evening twilight. Like Mercury, Venus is never found very far from the Sun in our sky. With a maximum elongation of 47°, its brilliance far outshines all other stars and planets in the sky. Only the Sun and Moon appear brighter.

From Earth, observers can watch as Venus goes from a thin crescent near inferior conjunction (see Figure 3.2) to a broad gibbous body just before or after superior conjunction. Galileo was the first person to record the changing phases of Venus. His observations

Figure 3.2
The brilliant planet Venus, as photographed by
the author. Exposure time was ¹/₁₅ second with
ISO 125 film through an 8-inch f/7 reflector
and 2× Barlow lens.

proved that the Sun is at the center of the Solar System, as theorized
by Nicholas Copernicus about a century earlier. Until Galileo's observations, it was assumed that every heavenly body revolved around
the Earth. According to the then-accepted Ptolemaic theory, if Venus
orbited Earth, then Galileo would have seen Venus go through a full
phase, just like the Moon (which does orbit Earth). With Venus
traveling instead around the Sun, full phase occurs at superior conjunction, when Venus is invisible.

Modern binoculars are capable of repeating Galileo's historic
findings. Venus measures about 1' of arc in diameter when it is closest
to Earth, which unfortunately coincides with inferior conjunction.
Though invisible in our sky due to glare from the Sun, Venus appears
only a little smaller for many weeks before and after conjunction,
during which time even 7× binoculars will reveal the thin crescent
phase.

Pulling farther away from the Sun as seen from Earth, Venus
slowly evolves from a crescent to a half-illuminated disk to a gibbous
body. All the while, the apparent diameter of the planet shrinks
progressively as the distance between our worlds grows. Observers
with binoculars will find it increasingly difficult to resolve Venus' disk
as the phases progress, especially when the planet nears superior
conjunction, its farthest distance from Earth.

On rare occasions, Venus may be seen to transmit the face of
the Sun at inferior conjunction. Whereas Mercury is seen as a perfectly
round disk at the time of transit, Venus appears slightly diffuse because of light refracting through its atmosphere. Telescopes of higher
magnification will also display an unusual "black drop" effect at
second and third contacts. The black drop effect results from sunlight

refracting through Venus' atmosphere. Venus appears to blend into the Sun's limb.

Transits of Venus occur in pairs separated by eight years, with over 100 years between successive pairs. The last transits of Venus were seen in 1874 and 1882. The next will take place on June 8, 2004, and June 5, 2012.

Mars

Named for the Roman god of war because of its blood-red color, Mars (Figure 3.3) has captivated mankind's imagination as no other planet has. Once thought to be the home of an advanced civilization, the Mars of today reveals many craters and sinuous channels seemingly formed by running water millions of years ago, as well as a huge dormant volcano named *Olympus Mons* and a 3,000-mile-long canyon named *Valles Marineris*.

From Earth, Mars never appears larger than 25″ of arc in diameter. Usually, it is even smaller than that. The best time to view Mars is when it is at opposition—that is, when the Earth-Mars distance is at a minimum and the planet appears largest. Table 3.1 lists upcoming oppositions of Mars for the remainder of the 20th century. Even then, however, binocular observers are at a disadvantage when looking at Mars. Apart from its distinctive red color, and perhaps just the slightest hint of the polar caps through high-power glasses, Mars reveals little.

Figure 3.3
Mars, as photographed on September 17, 1988, by Lee Coombs. He employed an 11-inch f/10 Schmidt-Cassegrain telescope for this ¼-second exposure at f/46 on Technical Pan 2415 film.

TABLE 3.1 Martian Oppositions: 1990–2000

Date of Opposition	Earth-Mars Distance (millions of miles)	Apparent Diameter (seconds of arc)	Apparent Magnitude
Nov. 27, 1990	48	18	-1.7
Jan. 7, 1993	58	15	-1.2
Feb. 12, 1995	62	14	-1.0
Mar. 17, 1997	60	14	-1.1
Apr. 24, 1999	53	16	-1.5

Jupiter

Mighty Jupiter, named for the king of the Roman gods, is also king of the planets. Measuring 88,000 miles in diameter, it surpasses in size all other bodies in the Solar System except for the Sun itself. Even at more than 391 million miles away, the immense size and high reflectivity of Jupiter cause it to appear brighter than any star in our night sky. At opposition, its great brilliance is overshadowed only by the Moon, Venus, and, at its brightest, Mars.

The highest power binoculars, such as 20 × 80s and 30 × 80s, reveal that Jupiter's clouds are segregated into distinct bright zones and dark belts, as shown in Figure 3.4. The bright zones are regions of gas rising to the cloud tops after being compressed and heated by the crushing pressure of the inner Jovian atmosphere. The darker belts are cooler gas cascading back down toward the interior. The *Great Red Spot,* a huge oval cyclonic storm found in Jupiter's southern hemisphere, is too small for even the highest power glasses to resolve.

The smallest glasses will show Jupiter's four largest moons: Io, Ganymede, Callisto, and Europa—the Galilean satellites. Discovered by Galileo, these satellites slowly but constantly change their positions relative to the planet and each other. For instance, two may appear on one side of Jupiter and two on the other, or perhaps three will be clumped together on one side and one on the other. Frequently, only two or three moons will be seen, with the "missing" satellite(s) either in front of or behind the Jovian disk.

It is all but impossible to tell which moon is which just by looking at them. Since the satellites change position on an hour-to-hour basis, it would take up too much time and space to show their relative locations every day for years to come. Fortunately, both *Sky and*

Figure 3.4
Jupiter with three of its Galilean moons. This photograph was taken by
Lee Coombs using an 11-inch f/10 Schmidt-Cassegrain telescope at f/46
and ISO 200 film. The planet was exposed for ½ second, while the
satellites required a five-second exposure.

Telescope and *Astronomy* provide this service to their readers in
every monthly issue. In addition, the Royal Astronomical Society of
Canada's *Observer's Handbook* and Guy Ottewell's *Astronomical
Calendar,* both annual publications, show the moon's daily positions.

Saturn

Saturn (Figure 3.5) is the most striking member of our planetary
family. With an equatorial diameter of about 75,000 miles, it is the
second largest planet circling the Sun. Appearing as a bright yellowish
starlike object in our sky, Saturn presents a maximum apparent diameter of 20″ and reaches magnitude −0.3 at opposition. This is about
ten times fainter than Jupiter.

    The planet's most attractive feature is, of course, the magnificent
ring system. Galileo noticed something odd about Saturn when he
viewed it through his first crude telescope in the early 1600s, but he
was unable to discern what exactly was odd about it. All Galileo saw
were two obscure protrusions on either side of the planet. The mystery

Figure 3.5
Saturn and its magnificent ring system. This
photograph, taken by James Fakatselis through
a 3½-inch Questar, was made using Technical
Pan 2415 film.

of Saturn remained unsolved for over half a century as other early telescopists tried in vain to decipher Galileo's observations. Finally, in 1665, a Dutch astronomer named Christian Huygens discovered the flattened ring encircling the planet.

Amateur astronomers who view Saturn through binoculars will experience the same frustration that Galileo must have felt nearly 400 years ago. Even when they are at maximum tilt toward the Earth, the rings cannot be resolved with less than about 35×. Giant glasses show Saturn as a football-shaped disk, but are unable to resolve the rings from the planet. Alas, it is sad but true that Saturn's beauty requires a telescope.

Circling Saturn are no fewer than 17 natural satellites. Titan, the largest Saturnian moon, appears as an 8th-magnitude "star" circling Saturn every 16 days. At greatest eastern or western elongation, it is 3.3′ of arc away from Saturn's disk. At those times, 7× binoculars are sufficient for spotting it, although some effort may be required. Binoculars of higher magnification will have little trouble showing Titan. Saturn's other major satellites are Mimas, Enceladus, Tethys, Dione, Rhea, Hyperion, Iapetus, and Phoebe. They are all quite faint by binocular standards and are never found far enough away from the planet to be readily distinguished in most glasses.

Uranus

Astronomical history was made on March 13, 1781. On that night, the renowned astronomer William Herschel accidently stumbled upon a round, greenish object in the constellation Gemini. At first, he mistakenly thought that he had discovered a distant comet, for this strange object moved very slowly against the background of fixed stars. Upon analysis of its projected solar orbit, however, it was determined that Herschel had discovered a new planet beyond Saturn's orbit. That planet later came to be called Uranus, the father of Saturn in mythology.

Uranus, the Solar System's third largest planet, measures 32,500 miles across. Unfortunately, our view of the planet has always been severely limited, given that it is over 19 times farther away from the Sun than Earth is. Voyager 2, after gliding past Jupiter in 1979 and Saturn in 1981, became the first (and thus far, only) spacecraft to fly past the cloud tops of Uranus in January, 1986. It revealed a world completely hidden by surprisingly bland greenish clouds of hydrogen and methane. Voyager also photographed the planet's tenuous ring system and its five icy moons, the only moons of Uranus known at that time. Perhaps the most impressive find of the Voyager encounter with Uranus was the discovery of ten additional satellites circling that distant world.

Through binoculars, Uranus is seen only as a greenish starry point (Figure 3.6). Telescopes are able to do little better: the greatest

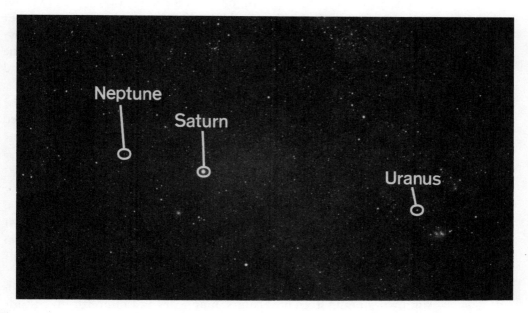

Figure 3.6
The planets Saturn, Uranus, and Neptune against the magnificent star fields of Sagittarius. This photograph was taken by Bernard Volz using a 135-mm f/2.8 lens on ISO 1000 film with a seven-minute exposure.

professional instruments show this distant world as a small, hazy, featureless disk. Annual finder charts for Uranus are published in the January issues of *Sky and Telescope* and *Astronomy*. Maps are also found in the Royal Astronomical Society of Canada's *Observers Handbook* and Guy Ottewell's *Astronomical Calendar*. These star maps will prove indispensable when you look for this elusive planet.

Neptune

By the beginning of the 19th century, astronomers had become mystified by the orbital motion of Uranus. Uranus was the only planet in the Solar System that did not behave according to the laws of planetary motion. Perhaps the irregularities astronomers found in its motion were caused by the gravitational pull from an unaccounted source—maybe another, unknown planet.

Using Newtonian mechanics, two astronomers, John Couch Adams of England and the Frenchman Urbain Leverrier, independently calculated how large the hypothetical world must have been, how far it was from the Sun, and where it was located in the sky. Leverrier sent positional information to the Berlin Observatory on September 23, 1846. Based on this information, Johann Gottfried Galle discovered Neptune later that same evening. Its location in the sky, near the star Mu Capricorni, coincided almost exactly with Leverrier's and Adams' predictions.

Our knowledge of Neptune has always been limited by the planet's tremendous distance away from Earth. All that changed forever in August, 1989, when the durable Voyager 2 flew by Neptune to reveal some amazing details. We now know that the planet is shrouded in a dynamic atmosphere of turquoise clouds. Voyager cameras revealed a dark earth-size vortex on Neptune that is reminiscent of Jupiter's Great Red Spot. Appropriately dubbed the Great Dark Spot, it appears surrounded by bright, wispy clouds. Voyager also revealed thin, irregular rings around the planet, as well as six new moons to add to Triton and Nereid, two satellites previously discovered through earthly telescopes.

Like Uranus, Neptune's challenge to amateur astronomers is simply in finding it. In binoculars, it appears as a faint 8th-magnitude "star" (Figure 3.6). Of course, there are so many stars of similar brightness, that confirming that one is observing Neptune can be difficult. Only its unique greenish color makes this planet stand out. Fortunately, annual finder charts for Neptune are published in the same periodicals as the charts for Uranus. Observers are strongly advised to use these charts when looking for either planet, as searching for them "unarmed" will only prove frustrating.

Pluto

Dubbed "Planet X" by astronomers of the late 1800s, distant Pluto was the subject of an extensive 25-year-long search before its existence could be confirmed. On February 18, 1930, a young astronomer named Clyde W. Tombaugh discovered Pluto using a specially designed camera at Lowell Observatory in Flagstaff, Arizona. Tombaugh was able to discern the planet only by its motion against the starry background, as Pluto appeared merely as a dim point of light on his photographic plates.

Even when Pluto lies at its minimum distance of 2.76 billion miles from the Sun, it is still an extremely faint object. With a brightest magnitude of only 13, it remains far below the detection threshold of binoculars.

4

Minor Members of the Solar System

Asteroids

On January 1, 1801, Giuseppe Piazzi was observing the sky from his home in Sicily when he happened upon a faint starlike object in the constellation Taurus. No doubt, he thought it strange that this star did not appear on his star charts of the region. He marked its position relative to the other, known stars and made it a point to return to that area on the next clear night. After a few days passed, he found, to both his surprise and delight, that his "star" had moved! Further observations permitted an orbit to be estimated for Piazzi's new object. It was found that whatever Piazzi had seen circled the Sun in 4.6 years and was 2.77 astronomical units (A.U.; 1 A.U. = 93 million miles) away. As is customary, the discoverer of a new member of the Solar System has the honor of naming it, and Piazzi chose "Ceres," after the Roman goddess of Sicily.

Of course, what Piazzi had discovered was an asteroid. Even when viewed through the largest telescopes on Earth, asteroids are seen as nothing more than starlike points of light, as Figures 4.1(a) and (b) demonstrate. In fact, that is why they were given the name *asteroid* in the first place: "aster-oid" means "star-like." The only way to identify an asteroid is to follow its movement against the more distant fixed stars.

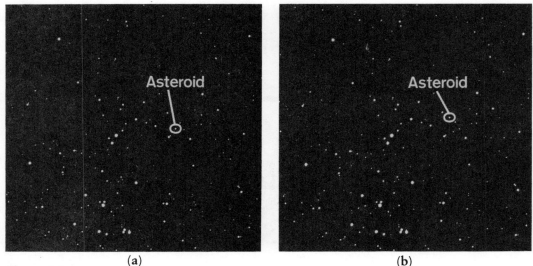

(a)   (b)

Figure 4.1
The asteroid (44) Nysa against the stars of the Beehive Cluster M44 in Cancer. Note the asteroid's motion in two nights' time. Both pictures were taken by Lee Coombs through a 610-mm f/6 lens on 103a-E spectroscopic film with an exposure time of ten minutes.

Over 60 asteroids become brighter than 10th magnitude and should therefore be visible through common binoculars. Table 4.1 lists the dozen brightest asteroids, along with their maximum magnitude and approximate size. Notice that (4) Vesta (the "(4)" signifies that it was the fourth one discovered) will occasionally even crack the naked eye barrier.

Looking for asteroids can be likened to hunting for a celestial needle in a haystack if proper star maps are not at hand. Whenever

TABLE 4.1 The Brightest Asteroids

Asteroid	Maximum Magnitude	Diameter (mi)	Asteroid	Maximum Magnitude	Diameter (mi)
(4) Vesta	5.1	345	(3) Juno	7.5	155
(2) Pallas	6.4	362	(18) Melpomene	7.5	102
(1) Ceres	6.7	637	(15) Eunomia	7.9	162
(7) Iris	6.7	138	(8) Flora	7.9	99
(433) Eros	6.8	12	(324) Bamberga	8.0	159
(6) Hebe	7.5	128	(1036) Ganymed	8.1	25

a minor planet of about 8th magnitude or brighter is due to be visible, astronomical periodicals will publish a finder chart with the object's path plotted against the stars. Look toward the area of sky where the asteroid is predicted to be. Once the area is centered, accurately sketch the stars visible in the field. While glancing back and forth between your field drawing and the asteroid's finder chart, carefully note which point of light on your map is not on the published diagram—that's the asteroid!

Serious "asteroiders" should consider joining the Association of Lunar and Planetary Observers' Minor Planet Section. One benefit of membership is receiving their publication *The Minor Planet Bulletin,* which keeps readers up to date on visible asteroids. To learn more about this organization, write Professor Frederick Pilcher, Department of Physics, Illinois College, Jacksonville, Illinois 62650.

An invaluable source of information on all the asteroids visible during a specific calendar year is the annual *Ephemerides of Minor Planets.* Produced by the Institute of Theoretical Astronomy in the Soviet Union, it offers sets of celestial coordinates for plotting each asteroid's path in the sky, the projected magnitudes of the asteroids, the dates of solar opposition, and other data. Copies of the annual *Ephemerides* may be obtained for a nominal fee from the Smithsonian Astrophysical Observatory, 60 Garden Street, Cambridge, Massachusetts 02138.

Comets

Of all the strange and beautiful celestial showpieces that appear in our sky, none have so captured our imagination as have comets (Figure 4.2). With their long, ghostly tails fanning out across the sky, they command our attention and awe whenever they flourish.

At any one time, there may be half a dozen or more comets in the sky. Most are visible only in large telescopes and pass unnoticed by most amateurs. At least once or twice each year, however, a new comet will break the magnitude 9-to-10 barrier and become bright enough to be seen in binoculars. The Smithsonian Astrophysical Observatory in Cambridge, Massachusetts, has been designated by the International Astronomical Union as the clearinghouse for announcing all discoveries and reappearances of comets. Astronomers around the world receive constant updates of any cometary activity through the *IAU Circulars,* IAU Telegrams, or the Central Bureau for Astronomical Telegrams' Computer Service. Details on subscribing to these services are available from the Smithsonian Astrophysical Observa-

Figure 4.2
Comet West, as photographed by George Viscome. The March 1976 picture was taken with ISO 400 film through a 50-mm lens and with an exposure time of ten minutes.

tory, 60 Garden Street, Cambridge, Massachusetts 02138. Information about the brighter, more interesting comets is also available from both *Astronomy* and *Sky and Telescope,* as well as through the recorded "Skyline" telephone service provided by the latter (617-497-4168; this is a toll call to Cambridge, Massachusetts).

Amateurs can make valuable contributions to the understanding of comets by making regular, systematic observations. Don't think you have to be "experienced" at this sort of thing; anyone can join in. One of the most useful features to note about a comet is its visual magnitude. By carefully following the changes in brightness of a comet and watching for any sudden fluctuations, we can closely monitor any physical changes within the nucleus.

The Morris method, introduced in the 1970s by renowned comet observer Charles Morris, is the most accurate way of estimating a comet's brightness. The observer begins by defocusing the comet *slightly,* until its bright center just begins to diffuse. The appearance of the comet is then memorized. Next, the comparison stars are defocused quickly, until they are similar in size to the retained image of the out-of-focus comet, and the magnitude of the comet is estimated. The observer repeats the observation as many times as is necessary to get a good value for the magnitude.

The Morris method requires locating appropriate comparison stars of known brightness. How can these be found? Frequently, a comet will pass through, or at least near, the field of a variable star (see Chapter 6). When that happens, many suitable stars may be found on the variable star's finder chart, which is available from the American Association of Variable Star Observers, 25 Birch Street, Cambridge, Massachusetts 02138. For a good selection of comparison stars across the entire sky, consider also the *AAVSO Star Atlas* (Sky Publishing Corporation, Cambridge, MA: 1980), an indispensable source when viewing comets that are brighter than 9th magnitude (the atlas' limiting magnitude).

Finally, the magnitudes of very bright comets must be estimated using just the naked eye. Some observers prefer to look through the "wrong" end of binoculars in order to shrink the comet down to a starlike point. However, finding bright stars for comparison may be difficult, since they are not always visible in the part of the sky the comet is traversing. In some instances, estimating by memory must suffice.

Another valuable observation to be made of a passing comet is assessing the diameter of its coma. The simplest way to determine this is to estimate the size of the coma with respect to two stars of known or measurable separation. It is also possible to judge the coma's diameter accurately just by knowing the field of view encompassed by the binoculars.

Watch to see if a tail develops behind a comet. Most comets visible in common binoculars should show at least a hint of one. If a tail does become visible, monitor its appearance, length, and position angle (P.A.) nightly. Measuring the tail's length is quite simple if you follow these steps:

1. Draw the field of stars in which the comet is situated.
2. Locate the position of the comet's head in relation to these stars.
3. Sketch the full length of the comet's tail, being careful to note just how far it extends.
4. Now go to a detailed large-scale star atlas (such as the *AAVSO Star Atlas* or *Uranometria 2000.0*) and match up the stars of your sketch.
5. Transfer the comet sketch onto the atlas. By knowing the scale of the map, the tail's length may be measured directly.

The tail's position angle is also easy to determine. Look at the comet and ask yourself, "Where is north?" Once north is established, if the tail is pointed in that direction, then its position angle is 0°. Of the other cardinal points, east is 90°, south is 180°, and west is 270°. If the tail lies exactly halfway between, say, south and west, then the P.A. would be 225°. If it is pointing more toward the west, it might be 240°, and so on.

Backyard astronomers have always been counted on to document closely the appearances of comets in our skies. Thanks to the dedicated patrol of amateurs from around the world, our knowledge of comets has continued to grow. The *International Comet Quarterly* acts as the international depository and clearinghouse for observations of comets. To subscribe or to inquire about submitting observations for publication, write to Daniel W.E. Green, *ICQ* Editor, Smithsonian Astrophysical Observatory, 60 Garden Street, Cambridge, Massachusetts 02138.

Comet Hunting

It is always fascinating to track a comet as it makes its way across the sky, but it is even more exciting to search for comets that have never been seen before. Imagine the thrill of being the first person in the world to spot one of these celestial visitors. You will become an instant celebrity in the global astronomical community, for the discoverer is always honored by having his or her name attached to the comet.

Don't be misled into thinking that just anyone can be a comet hunter. While a comet may certainly be found by accident, most are discovered by dedicated amateur astronomers who have already spent hundreds, even thousands, of fruitless hours scanning the skies. Many hours of sleep must be forsaken when stalking comets, but it will all be worth it in the end when that first one is found!

Binoculars (especially giant glasses) lend themselves nicely to comet hunting. The glasses should be coupled to a steady mount that permits smooth, easy sweeping in both altitude and azimuth. Special attention should be paid to the comfort of the observer when choosing a mount. The ideal binocular setup should have the user seated and the eyepieces at or near eye level. Searching in either a standing or bent-over position, or with hand-held binoculars, will only serve to interfere with an observer's concentration. How can anyone look for a faint smudge of light when his or her back is aching?

Since comets are at their brightest near perihelion, most successful hunters search the western horizon in the early evening and the

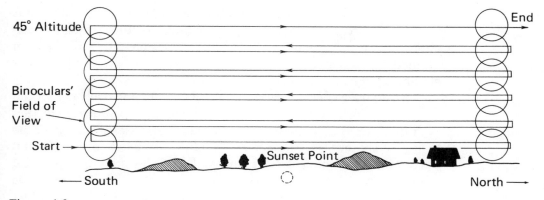

Figure 4.3
Search pattern recommended for comet hunting shortly after sunset. Illustration by the author.

eastern horizon in the early morning. Figure 4.3 shows a good hunting strategy. For the evening check, begin at the horizon and slowly scan a wide breadth on either side of the sunset point. Many sweep all the way from due north to due south. At the end of the pass, swing back to the starting point, move the instrument up half a field diameter so as to overlap part of the view, and scan back across an equal width of sky. Continue this motion until you have reached an altitude of between 45° and 60° above the horizon.

The morning check is performed in exactly the reverse order. Start high in the eastern sky, and scan across and down toward the horizon. Statistics show that, for some unknown reason, nearly three times as many comets are discovered in the morning sky as in the evening sky.

The hearts of many new comet hunters skip a beat when they come upon an unexpected fuzzy smudge on their very first night out. Can comet hunting be this easy? Ah, if only it were. There are hundreds of deep sky objects within the range of binoculars that look like comets at first. A quick check of a detailed star atlas should squelch the excitement when it is found that the new "comet" is actually a distant cluster, nebula, or galaxy that has been known for years!

If no corresponding deep sky object is found on an atlas, then check your mystery object for motion. Comets move in the sky; deep sky objects do not! Plot the suspect accurately on your star atlas, and wait for some sign of movement. Comets move quite slowly, so be patient. If many hours go by and there is still no change, you have probably spotted something else, but don't give up that easily. Go back out the next night and check again. If the object still has not moved, then you have not found a comet.

But what if it *did* move? The first thing to do is to check your available sources, such as the *IAU Circulars* or the "Skyline" telephone message, to make sure that the comet has not already been claimed by somebody else, or that it is not just a periodic comet making a scheduled return.

If you have checked everything and you are positively convinced that this is not a previously discovered comet, an unidentified deep sky object, or even a closely set asterism of faint stars, then it is time to notify the authorities. Send a telegram or telex to the Central Bureau for Astronomical Telegrams at the Smithsonian Astrophysical Observatory (SAO) in Cambridge, Massachusetts. The SAO may be reached through their Western Union telex number, 710-320-6842 (answerback ASTROGRAM CAM). In the message, the following must be included:

1. Your name, address, and telephone number.
2. The date and time (in U.T.) of the observation in decimals of a day.
3. The right ascension and declination of the suspected comet.
4. The direction of motion of the object.
5. The magnitude of the object.

With the message received, observatories around the globe are put on alert to confirm your suspected comet's existence. Even after all of your checking, it still might be a false alarm, or another observer may beat you to the punch. With any luck though, yours might be the first report received at the SAO, thereby adding your name to the distinguished list of comet discoverers. Good luck!

Meteors

Observing meteors (Figure 4.4) through binoculars is an area of amateur astronomy that has remained unexplored by most hobbyists. With the naked eye, only those meteors brighter than 4th magnitude are readily detectable. Through 7× binoculars, however, the range is expanded to 7th, 8th, or even 9th magnitude. The number of meteors capable of being seen is therefore greatly increased.

Unfortunately, along with this greater range of magnitudes comes a markedly reduced field of view. Wide-angle glasses prove to be the instrument of choice, since we want to be able to view the largest portion of sky possible. Both 7 × 35 and 7 × 50 binoculars

Figure 4.4
A bright Perseid meteor blazes across Andromeda in August 1983.
Photograph by Lee Coombs using a 50-mm f/2 lens and ISO 400 film.

offer up to a 10° to 12° field. Even though edge definition sometimes suffers, these would seem the best compromise between magnification and field size.

Observing meteors with binoculars is one of the few areas of amateur astronomy left where dedicated observers can make a valuable contribution. Karl Simmons of the American Meteor Society suggests two methods for amateurs to become involved in this activity. The easiest way is simply to count the number of meteors seen. Keep your sights fixed on a specific area of the sky, tracking the stars as the Earth rotates. Record how many meteors were seen, and be sure to include the area of sky monitored and the time (U.T. again) that observations began and ended.

A second, more sophisticated approach is to record several facts about each meteor seen, such as the date and time of the meteor's appearance, the magnitude of the meteor (to the nearest half-magnitude), and the duration and direction of travel of the meteor. The *AAVSO Star Atlas* is an excellent source of comparison stars to use in determining the magnitude and is also useful for plotting a meteor's extent of travel.

The duration of a meteor's passage, in tenths of a second, may be estimated as follows: 0.1 to 0.2 for a very swift meteor, 0.3 to 0.6 for a meteor of average speed, and 0.6 or more for a slow meteor such as a bolide. Include in your observations a description of any persisting smoke trail, noting both its duration and direction of drift. For the observations to be of value, state the type of equipment used, the field of view in degrees, and the position in the sky (right ascension and declination) at which your binoculars were aimed.

Table 4.2 lists the dates of the chief meteor showers that occur each year, together with pertinent information about them.

The American Meteor Society welcomes amateurs who wish to contribute to this little-known field. Standard report forms, as well as a paper by Simmons titled "Telescopic Meteor Observation," are available from the American Meteor Society, Department of Physics, State University of New York, Geneseo, New York 14454.

TABLE 4.2 Major Annual Meteor Showers

Date of Maximum	Shower Name	Location of Radiant R.A.	Dec.	Average Hourly Rate*	Associated Comet
January 4	Quadrantids	15h28m	+50°	40>	?
April 21	Lyrids	18 16	+34	15	Thatcher
May 4	Eta Aquarids	22 24	0	20	Halley
July 28	Delta Aquarids	22 36	−17	20	?
August 12	Perseids	03 04	+58	60>	Swift-Tuttle
October 21	Orionids	06 20	+15	25	Halley
November 3	South Taurids	03 32	+14	<15	Encke
November 17	Leonids	10 08	+22	15	Tempel
December 14	Geminids	07 32	+32	50>	?
December 23	Ursids	14 28	+76	15	Tuttle

* For naked eye meteors only.

5

The Sun

Most of us restrict our sky watching to the hours between the end of evening twilight and dawn's first light, for it is during these dark hours that the stars, Moon, and planets are seen. But by doing so, we are ignoring the most important star in the sky—the Sun.

Before we begin our journey to the Sun, it is critically important to exercise extreme care. Without proper precautions, you may suffer permanent eye damage and even blindness. If you are unsure of how to view the Sun, take a moment right now to review the section on safety during solar observation further along in this chapter.

Many amateur astronomers enjoy monitoring the constantly changing face of our star. For most, its principal attraction is the fluctuating number of sunspots across the *photosphere,* or surface, of the Sun. *Sunspots* (Figure 5.1) are transient features that apparently result from tears in the photosphere. These rips are associated with powerful magnetic fields that block some of the outward-pouring radiation from the Sun's core. The resulting temperature of the affected region is about 3,000°F cooler than the surrounding photosphere, causing the sunspot to appear darker by contrast. Each sunspot consists of a black central portion known as the *umbra* and an encircling grayish ring called the *penumbra*. Sunspots may range in size from hundreds to thousands of miles in diameter.

Typically, sunspots require some magnification to be seen, although the largest may be spotted with the naked eye provided that

Figure 5.1
The Sun's active photosphere alive with many sunspots. Brian Kennedy took this photograph at the prime focus of an 8-inch f/5 telescope, exposing ISO 125 film for 1/250 second.

proper eye protection is used. How many sunspots are visible at any one time depends on a number of factors. Naturally, higher magnification will resolve smaller spots. However, the number of visible sunspots also depends on where the Sun is in its current sunspot cycle. Astronomers in the 19th century determined that sunspot numbers oscillate regularly over approximately an 11-year period. Toward the maximum phase of the cycle, there might be a hundred or so visible spots; at minimum, the solar disk will be nearly devoid of them. The most recent solar maximum occurred in early 1990; the next is expected in about 2001.

At or near a sunspot cycle maximum, the frequency of solar flares also increases. *Solar flares* are colossal eruptions associated with larger sunspot groups. Usually, flares require advanced narrow-band filters to be seen, although extremely rare "white-light" flares have been observed.

While flares may not be readily observable, their effect may be detected on Earth. As a flare erupts, charged particles are emitted from deep within the Sun's interior. Racing outward at harrowing speeds of over 500 miles per second, these particles reach Earth's orbit in about two days. As they bombard the Earth, they are drawn toward our planet's magnetic poles to produce the beautiful aurorae—the northern lights in the northern hemisphere and the southern lights in the southern hemisphere. Regardless of their intensity, aurorae are best viewed with just the unaided eye: the majesty of the event will be completely lost in even the widest angle binoculars, which have too restricted a field of view.

Observers using high-power glasses will also be able to detect *faculae,* brighter areas set close to major sunspots. Faculae give the impression of being brighter than the surrounding photosphere, but that is only because they lie a couple of hundred miles above the visible surface. At that altitude, they are not as affected by the solar atmosphere's dimming effect as is the visible photosphere itself. Faculae are most easily seen near the Sun's limb. Here, the contrast between faculae and photosphere is at its greatest, thanks to *limb darkening,* the effect produced when we look toward the disk's edge rather than its central region and thus view the Sun diagonally through a deeper area of its atmosphere. Observers should note that to spot faculae requires at least $15\times$ to $20\times$ glasses.

Safety First!

Binoculars are fine instruments for monitoring the Sun's fluctuating sunspot activity. However, before looking at the Sun, take warning:

DO NOT LOOK DIRECTLY AT THE SUN, NOT EVEN FOR AN INSTANT!

The Sun's ultraviolet rays, the same rays that cause sunburn, will burn the retinas of your eyes much faster than your skin. Without proper safety precautions, permanent eye damage, and even blindness, can result.

There are several ways to look at the Sun safely. If you wish to view the Sun directly through binoculars, special solar filters are required. *Do not use welder's glass, smoked glass, or overexposed film:* they all can cause blindness. Proper filters are commonly made of aluminized Mylar™ film or glass and must be securely mounted *in front* of both objective lenses beforehand. In this way, the dangerously intense solar rays are reduced to a safe level prior to entering the binoculars and your eyes.

Never place the filters between the eyepieces and your eyes. Binoculars will magnify not only the Sun's light, but its intense heat as well. The concentrated solar heat can quickly crack glass filters or burn through thin Mylar film, allowing undiminished sunlight to burst into your unprotected eyes and possibly cause permanent blindness.

Solar filters must be treated with great care, or they will quickly become damaged and unsafe to use. Inspect the filters regularly (especially the Mylar type) for pinholes or irregularities in the coating by holding the filter up to a bright light. If a pinhole develops, it need not be cause to discard the filter: small holes may be sealed with a tiny dot of flat black paint without compromising the image too greatly. Using a cotton swab, dab just a bit of paint over the hole. If more serious damage than a pinhole is detected, then the filters must be replaced immediately.

The safest way to view the Sun is to use your binoculars to project an image of the Sun onto a piece of smooth, clean, white cardboard (Figure 5.2); any irregularities in the projection screen will cause the image quality and amount of visible detail to suffer. First,

Figure 5.2
The author's daughter, Helen, demonstrates the safest way of observing the Sun with binoculars. Note the cardboard sun shade. Photograph by the author.

mount the glasses on a tripod or other rigid support. Keep a dust cap over one of the objective lenses to prevent two overlapping images. *Do not look through the binoculars at any time, not even to aim them.* You can align the glasses by adjusting the shadows they cast until they are at their shortest.

Once the binoculars are centered on the Sun, focus the lens until the image is sharp and clear. If the image is not bright enough to be seen well, cut out a cardboard baffle that fits securely over the front of the glasses' barrels, as shown in the figure. In this way, unwanted sunlight will be blocked from reaching the projection screen and washing out the solar image.

This projection method lends itself very nicely to keeping a permanent record of sunspot activity. Depending on the size of binoculars used, draw a circle three to four inches in diameter, in the center of a piece of paper, and attach it to the projection screen. Adjust the distance between the binoculars and the screen so that the projected Sun fills the circle exactly. Once this is done, simply trace the exact positions and sizes of sunspots. Be sure to note the four cardinal directions on your observation sheet, and remember that over a period of days, sunspots appear to travel from west to east across the Sun.

Another interesting activity is to draw the solar disk every second or third day over a span of a month. With each successive sketch, sunspots will be observed to form and change in size and shape (some quite radically) as they travel across the solar disk. From this exercise, you can directly estimate the Sun's period of rotation in much the same way as Galileo did over 350 years ago. By paying special attention to the exact location of the visible spots relative to each other, it will soon become clear that different latitudes on the Sun rotate at different speeds. For instance, the Sun's equator rotates once every 25 days, while the polar regions take more than a month to complete one turn.

Solar Eclipses

Solar eclipses always attract a lot of attention from amateur astronomers, as well as from the general public. Enthusiasts around the world await eagerly their precious time in the shadow of the Moon.

A total solar eclipse (Figure 5.3) may be the most magnificent celestial event of all. Solar eclipses occur as the Moon moves across the face of the Sun, obscuring our view of the solar disk. Excitement mounts when about 85% of the Sun is covered and the sky begins to darken. The temperature falls, and winds may pick up. Animals are

Figure 5.3
A total eclipse of the Sun. Jack Newton took this hand-held ¹/₆₀-second exposure through the window of an airplane using a 200-mm f/4 lens and ISO 400 film.

fooled into thinking that night is approaching, and birds may go to roost. The brighter planets and stars become visible as the ever-deepening twilight encroaches.

In the few seconds before totality, the Moon's jagged edge breaks up the last remaining rays of sunlight into brilliant luminous fragments. These were first described by Francis Baily, an English astronomer, during an 1836 eclipse. He likened their appearance to a dazzling necklace of beads, and so today we refer to the phenomenon as *Baily's Beads*. Frequently, one bead will outshine all others, creating the *diamond-ring effect*.

At *second contact,* the Moon completely covers the Sun. It is now safe to view the Sun without protection, since the dangerous rays of the photosphere are blocked. As the last few beads of sunlight disappear, the beautiful *corona,* or outermost atmosphere, of the Sun blossoms into view. Binoculars will reveal its pearly white glow spreading out for several degrees from the Sun. The corona will remain visible only during the period of totality.

The *chromosphere,* or inner atmosphere, of the Sun will also pop into view at second contact. Steadily mounted binoculars display beautifully the reddish glow encircling the Moon's blackened limb. Within the chromosphere are the spectacular prominences, shooting upwards into the corona, only to loop back if their velocity is not

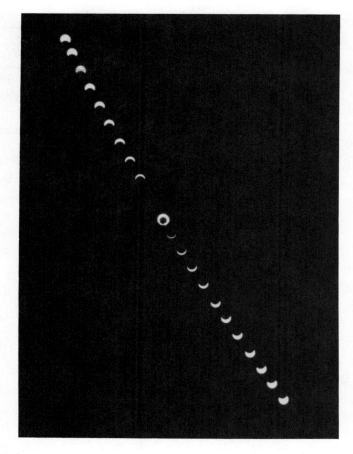

Figure 5.4
The unusual annular-total solar eclipse of May 30, 1984, as
photographed by the author from Lake Norman, North Carolina.
Exposures were made at five-minute intervals (oops, missed one!) by
exposing ISO 64 film at ¹/₁₂₅ second with a 50-mm lens set at f/11 and
shielded with a neutral density 5 filter.

great enough to escape the Sun's tremendous gravity. High-power
glasses are best to view these magnificent eruptions.

The minutes of totality will pass faster than any you may have
ever experienced. *Third contact* is reached when the Moon's disk
slides off the Sun. Once again, the brilliant diamond-ring effect and
Baily's Beads are visible.

Total eclipses occur because the Sun and Moon appear to be
almost exactly the same size in our sky. However, since neither the
Moon's orbit around the Earth, nor the Earth's orbit about the Sun,

TABLE 5.1 Solar Eclipses, 1990–2000

Date (U.T.)	Type of Eclipse	Zone of Maximum Eclipse
22 July 1990	Total	Finland, northernmost Soviet Union, Aleutian Islands, northern Pacific Ocean
11 July 1991	Total	Pacific Ocean, Hawaii, Baja California, mainland Mexico, Central America, Colombia, Brazil
4–5 January 1992	Annular	Pacific Ocean
10 May 1994	Annular	Pacific Ocean, northwestern Mexico, North America (from Texas through New England and into Ontario and the Maritime Provinces)
26 February 1998	Total	Pacific Ocean, Colombia, Venezuela, Atlantic Ocean
11 August 1999	Total	Atlantic Ocean, England, France, Germany, Austria, Hungary, Romania, Turkey, Iran, Pakistan, India

is exactly circular, the apparent size of the Moon and Sun can vary slightly. When the Moon appears a little smaller than the Sun at the time of a total solar eclipse, a bright ring, or *annulus*, of sunlight remains visible during the maximum stage of the eclipse. Unlike true total eclipses, *annular* eclipses (Figure 5.4) require constant eye protection to be viewed safely, even at maximum phase. Likewise, full precautions must be exercised during all stages of a partial solar eclipse.

Table 5.1 lists contemporary annular and total solar eclipses through 1999. If you plan on viewing one of these magnificent events, you are urged to review the monthly and annual astronomical periodicals well beforehand to learn the specific details.

6

Stellar Happenings

Whether you use a pair of 7 × 35 binoculars or 25 × 105 giant glasses, the night sky will explode with treasures that remain unsuspected to the naked eye. Some seemingly mundane stars burst with color when viewed through binoculars. Numerous double stars dot the heavens, beckoning us to take a peek. Variable stars mysteriously fluctuate in brightness; at times, some may be seen with the naked eye, while at other times, their dim light may barely register through binoculars. The list seems almost endless, with so many fine clusters of stars, huge interstellar clouds of gas and dust, and remote galaxies to choose from.

Double and Multiple Stars

Nearly half of the stars we see at night are double or multiple star systems (Figure 6.1). Of course, not all can be resolved through binoculars, but of the hundreds that can, no two appear exactly the same. Many display striking contrasts in magnitude, while others are nearly equal in brightness. Some seem to shine pure white, while others glimmer with distinctive colors. The separation between the stars varies from resolvable with the naked eye to detectable by spectral analysis only.

Figure 6.1
Albireo (Beta Cygni), a superb double star.
Photograph by Lee Coombs at the prime focus
of his 10-inch f/5 Newtonian reflector with an
exposure of 15 seconds on ISO 125 film.

There are three types of double stars that are visible in binoculars. *Visual binaries* are two or more stars that actually orbit each other in space. The brightest star in such a binary system is typically referred to as *A*, while the fainter secondary member is labeled *B*. If there are other, still fainter components, they are specified as *C, D,* and so forth. *Optical doubles* are simply chance alignments of two or more stars that lie along the same line of sight from Earth, but are actually nowhere near each other in space. The third class of double star is the *eclipsing binary;* we shall discuss this kind of system in the section on variable stars.

Double stars are always referred to by their separation and position angle. A binary's *separation* is simply the apparent angular distance between the primary *A* star and the secondary *B* star and is usually expressed in seconds of arc. Clearly, the tighter the stellar pair, the higher the power and the larger the aperture needed to resolve them. All of the double stars listed throughout Chapter 7 are resolvable in 7× to 10× binoculars, unless otherwise noted.

The *position angle*, or P.A., is the angular measurement of an imaginary line drawn from the primary to the secondary relative to north. If the companion is due north of the primary, then the pair's P.A. is 0°. If the companion is due east, then the P.A. is 90°. Similarly, a P.A. of 180° places the companion due south, while a P.A. of 270° indicates that the companion is due west of the primary.

Keep in mind that a binary system's separation and position angle change over time as the stars complete their orbits. The values for each double star listed in the next chapter are taken from modern authorities, yet some may have changed since publication. In an effort to clarify this, each listing specifies the year in which the measurements were made. The separation is listed in its own column, while the position angle is specified under the "Comments" column. As an example, consider Beta Cygni. The data table (page 137) lists a component separation of 34″ of arc, while the remarks for this entry read: "54° (1967); 12540; Albireo." This may be interpreted as indicating a separation of Beta Cygni's components of 34″ with the secondary star found at P.A. 54° as measured in 1967. The second number is the star's listing in Aitken's *New General Catalogue of Double Stars*, while the final term is the star's proper name, Albireo.

Variable Stars

Variable stars [Figures 6.2(a) and (b)] are distant suns that appear to fluctuate in brightness. Some do so rhythmically over a predictable period of time, while others vary irregularly. Many variable stars suddenly flare in a matter of minutes or hours. There are three major classes of variables: *eclipsing binaries, pulsating stars,* and *eruptive variables.*

Strictly by chance, some binary star systems are seen nearly edge on from our earthbound vantage point. Called *eclipsing binaries,* these stars alternately pass in front of each other, causing temporary diminishings of their combined light. In general, a steep drop in magnitude is seen when the smaller *B* companion transits in front of the primary *A* sun, while only a slight fading occurs when the secondary is eclipsed by the main star.

One especially interesting eclipsing binary to view through binoculars is Algol, in autumn's Perseus. By comparing it to other stars of known brightness, we find that, over a period of about three days, Algol changes from 2nd to 3rd magnitude and back again. It was because of this seemingly possessed behavior that, long before the invention of the telescope, our ancestors named the star "Algol," the Arabic word for "demon."

Eight subgroups of variable stars are categorized as *pulsating stars*. These stellar oddities actually experience physical expansion and contraction as we watch them alternately brighten and fade. The most common type is the *long-period* class of stars. Over a period of weeks or months, these red giant suns are observed to fade and brighten with precise regularity. Long-period variables are frequently called *Mira-type* stars, after the star Mira, which was the first long-period variable discovered. Located in the autumn constellation Cetus

Figure 6.2
The classic long period variable Mira (Omicron Ceti) near maximum (**a**) and minimum (**b**) brightness. Both photographs were taken by Lee Coombs using a 150-mm f/4 lens and exposing ISO 400 film for five minutes.

the Whale, Mira is easily visible with the unaided eye near maximum brightness, but fades to near invisibility through binoculars at minimum.

Semi-regular variables are also identified as red giants, but frequently experience random fluctuations in their regular cycles. *Irregular variables* change in magnitude over an unpredictable length of time. They are especially fun to watch night after night, as you never know what to expect.

RR Lyrae-type stars, named after RR Lyrae, the first of the type to be discovered, have very short, precise periods of one day or less. Most are found in globular clusters and thus are also called *cluster variables*.

Cepheid variables also experience brightening and fading over short lengths of time, usually one to seven days. Named after their prototype star Delta Cephei, Cepheids are intrinsically very bright giant stars that are noted for their *luminosity-versus-brightness* relationship. It is well known that the length of a Cepheid's variability period is directly proportional to its absolute magnitude, a fact which makes Cepheids quite valuable for determining distances in the universe.

RV Tauri stars are pulsating orange and yellow supergiant suns that exhibit alternately shallow and deep minima. The length of time between two primary minima is typically between about 30 to 150 days.

Z Andromedae variable stars (also known as *symbiotic variables*) are close binary systems whose spectra seem to suggest a cool giant sun paired with a hot companion. The unpredictable periods of variability of symbiotic stars probably result from a combination of pulsations by the larger component and physical interaction between the two stars.

S Doradus stars are among the most luminous known. These young, massive blue stars undulate an average of three magnitudes over either rhythmic or irregular periods. Analysis of their spectra indicates these brightness deviations are caused by high-speed ejection of the star's atmosphere.

Eruptive variables form the third major class of variable stars and, like pulsating stars, may be further broken up into several branches based on unique characteristics.

Novae are white-hot dwarf stars that suddenly brighten by five magnitudes or more in only a few days and then fade back to their original luminosity over the next few months or years. *Recurrent novae* are stars that have been observed to repeat the nova cycle two or more times.

If a star is much more massive than the Sun, the end of its life will be signaled by a tremendous explosion. During such a *supernova* outburst, nearly all of the star's mass is instantly converted into energy. For a short while, the brilliant supernova may outshine all of the billions of other stars in its home galaxy combined.

Supernovae are rare events. The last one seen within the Milky Way occurred in 1604, only five years before the invention of the telescope, and there hasn't been another one since. The odds say that we are long overdue. In the meantime, we will have to settle for spotting supernovae in distant galaxies.

Much excitement was generated in February 1987, when a supernova flared in the Large Magellanic Cloud, one of the closest galaxies to our Milky Way. The explosion gave astronomers the unprecedented opportunity to study a supernova from "only" 163,000 light years away. At maximum, the supernova reached magnitude +3, which translates to an absolute magnitude of −15.5!

U Geminorum or *Dwarf Novae* stars are similar to recurrent novae, but flare repeatedly at roughly two-month intervals. These suns burst forth suddenly in a day and subside to their normal brightness within a week.

R Coronae Borealis stars are normally quite bright, but fade away quickly at irregular intervals. Then they begin the slow ascent back up to their original magnitude over the next few weeks or months. These "celestial Cheshire cats" have been likened by many authorities to novae in reverse.

Gamma Cassiopeiae stars are thought to be rapidly spinning spectral type-B stellar infernos that exhibit random magnitude fluctuations of up to about 1½ magnitudes.

Flare stars are red dwarfs that, without warning, burst forth and rise several magnitudes in a minute's time. Constant around-the-clock monitoring is required to study the flaring of these most unusual suns.

Star Clusters and Associations

Scattered throughout the sky are loosely bound throngs of hot, new stars of spectral classes O (blue) and B (white). These groups, all of whose stars formed from the same interstellar cloud, are called *OB associations*. OB associations are fleeting stellar affiliations: the bonding force of gravity is usually sufficient to contain the group for only a few million years, after which time the stars begin to go their separate ways.

While there are about 70 OB associations known, few are visually distinguishable. With their constituents so loosely packed, it can be difficult to tell visually which are true members and which just happen to lie in the same direction but are nearer or farther away.

Like OB associations, *open clusters* (Figure 6.3) are randomly shaped collections of primarily young blue and white suns, but they are more permanent heavenly fixtures. Also known as *galactic clusters,* each may contain up to a few hundred individual points of light. Some, such as the Hyades in Taurus or the Coma Berenices cluster,

Figure 6.3
M67 (NGC 2682) in Cancer, a stunning though often-ignored open cluster seen in the spring. This photograph by George Viscome was taken through a 500-mm lens with ISO 400 film exposed for ten minutes.

are loosely grouped, while others, such as M11 in Scutum, are crowded into stellar traffic jams.

Many fine open clusters may be seen with binoculars. Some are ablaze with colorful suns, while others appear as little more than hazy smudges. Many clusters listed in the next chapter have apparent diameters in excess of one degree, making them difficult to detect in telescopes with narrow fields. Since most observational handbooks cater to the telescopist, the existence of many of these clusters have been almost totally ignored. Binoculars, on the other hand, are able to distinguish them fairly readily from the starry background. It is hoped that their mention here will alert observers to their existence. Do not pass them by, as many are quite striking.

Surrounding the center of our galaxy, much as moths gather around the flicker of a flame, are huge, spherical conglomerations of stars—the magnificent *globular clusters* (Figure 6.4). Sir William Herschel was the first astronomer to identify the true nature of these massive global conglomerations. He coined the phrase "globular cluster" for the first time in 1786, when he described some of the entries in his catalogue.

Even at the vast distances at which they lie, globular clusters appear surprisingly bright. Of the more than 100 known, nearly half

Figure 6.4
The magnificent globular cluster M13 (NGC 6205), seen in Hercules.
George Viscome took this photograph through a 14½-inch Newtonian
reflector using a cold camera loaded with ISO 400 film. Exposure time
was 30 minutes.

are visible in binoculars. Unfortunately, due to their tight stellar
packing, resolution of individual stars will prove impossible in most
cases. A noteworthy exception is Omega Centauri (NGC 5139), the
most magnificent of its kind. Found in the southern constellation
Centaurus, this breathtaking sphere of stars is partially resolvable in
7× glasses.

Nebulae

Drifting amidst the vacuum of deep space are huge clouds of gas and
dust called *nebulae*. Searching for these vaporous celestial phantoms
is both exciting and frustrating.

Stars are born from *diffuse nebulae* (Figure 6.5). These great
clouds are found primarily in the outer regions of the Milky Way
and in similar galaxies. Diffuse nebulae are composed primarily of
hydrogen, helium, nitrogen, and oxygen. There are three types: emission, reflection, and dark nebulae.

Figure 6.5
M16 (NGC 6611), a fine example of a diffuse nebula, is found in the summer constellation Serpens. Photograph by Kim Zussman.

As stars form within diffuse nebulae, the component gases enter a state of high molecular excitation, causing the nebulae to glow, much like the neon gas in a sign. These "star factories" are referred to as *emission nebulae*. The Great Nebula in Orion (M42) is the finest example of an emission nebula that is easily visible from the northern hemisphere. Embedded within its hot clouds are several stars that have "turned on" only within the past 100,000 years or so.

Eventually, the mass of an emission nebula will be insufficient to form additional suns. Now called *reflection nebulae,* they remain visible only by reflecting light from the stars they surround. Over time, the remaining clouds will totally dissipate due to radiation pressure from those same stars, originally begotten from the clouds. Winter's M78 (NGC 2068), in Orion the Hunter, is a good example of a reflection nebula.

The most elusive diffuse nebulae are actually invisible. Instead of emitting or reflecting light like their brighter counterparts, these *dark nebulae* (Figure 6.6) obscure light from the stars behind them. They appear as starless holes silhouetted against a star-filled field. The best known example of a dark nebula is the Horsehead Nebula in Orion, although it is too small and faint to be visible in binoculars. However, many other fine examples will show up.

Just as diffuse nebulae are associated primarily with stellar birth, other nebulae are associated with stellar death. Contrary to their name, *planetary nebulae* (Figure 6.7) have no relationship to the Solar System. They are thought to originate from the gases that are expelled by novae. As these gases expand, bright spheres surrounding the erupting stars are formed. The spherical shells of gas are composed largely of ionized oxygen, causing most planetaries to glow green

Figure 6.6
Everyone's favorite dark nebula, Barnard 33 in Orion, is better known by its nickname, the Horsehead Nebula. Photograph by Jack Newton.

Figure 6.7
NGC 1501 is an out-of-the-way planetary nebula in Camelopardalis that is visible in binoculars as a starlike point. Photograph by Lee Coombs through a 10-inch f/5 Newtonian. Exposure time was five minutes on 103a-E spectroscopic film.

or turquoise. The Dumbbell Nebula (M27), found in the summer constellation Vulpecula, is the most distinctive planetary visible through binoculars.

Supernova remnants (Figure 6.8) result from the deaths of the most massive stars known. In the year 1054 A.D., Chinese astronomers recorded the appearance of a "guest star" near the eastern horn of Taurus the Bull. It became so bright that it was easily seen in the daytime sky for several weeks. Finally, two years after it first became visible, this mysterious star vanished. When today's telescopes are aimed in the same direction, all we find is an expanding cloud of interstellar leftovers, with a rapidly spinning pulsar buried deep within. We know it as the Crab Nebula (M1) because of its unusual photographic resemblance to a crab.

Galaxies

A galaxy may be thought of as a continent in the infinite ocean of the universe. Each is made up of hundreds of millions or even billions of individual stars. The Milky Way is but one of millions of known galaxies in the universe.

Figure 6.8
The most famous supernova remnant is undoubtedly the Crab Nebula M1 (NGC 1952) in Taurus. Photograph by Martin Germano using an 8-inch f/10 Schmidt-Cassegrain telescope for this 75-minute exposure on hypered Technical Pan 2415 film.

Until the application of telescopic photography, galaxies were thought to be no more than nebulous patches. From extensive studies of photographic plates taken in the early part of the 20th century, the astronomer Edwin Hubble concluded that galaxies are actually vast independent stellar systems. He demonstrated further that all galaxies fit into three major classes according to their physical shape.

Many galaxies resemble huge pinwheels and are thus termed *spiral* galaxies (Figure 6.9). Each spiral galaxy consists of a bright central nucleus, where older stars are found, and several arms unwinding outward. Most spirals have two arms, where young stars, open clusters, and diffuse nebulae are located. The Milky Way is known to be a spiral galaxy, with the Sun located about two-thirds of the way out from the galactic center.

Spiral galaxies may be broken down into five subclasses, according to how tightly the arms are wound around the nucleus. Galaxies with the tightest arms are *Sa* spirals, those with moderately tight arms, such as the Milky Way, are *Sb* galaxies, and loosely coiled spirals are classified as *Sc*. *Sd* spirals are comparatively rare galaxies with

Figure 6.9
M81 (NGC 3031), in Ursa Major, is one of the most stunning spiral galaxies in the entire sky. Photograph by Kim Zussman.

extremely wide arms. Just as unique are the *S0* galaxies, which Hubble suggested may be transitional objects between the spiral and elliptical classifications.

Further study revealed that some spirals exhibit an odd, barlike feature passing through the nucleus. The spiral arms of these *barred spirals* begin to unwind from the ends of the bar, rather than from the nucleus itself. Like normal spirals, barred spirals are classified according to the tightness of the arms, with designations *SBa*, *SBb*, and *SBc* applying as above.

Figure 6.10
Virgo's M84 (NGC 4374) and M86 (NGC 4406), a pair of elliptical galaxies, are the two brightest of many distant galaxies found in this photograph by Lee Coombs. Exposure time was 15 minutes at the prime focus of a 10-inch f/5 Newtonian reflector using 103a-O spectroscopic film.

Elliptical galaxies (Figure 6.10) are the most plentiful type of galaxy in the universe. In an elliptical, there is not even a hint of spiral arms. In fact, ellipticals appear as huge, oval spheres with no internal structure of any sort. They are classified along a seven-point continuum according to how out of round they appear. For instance, *E0* ellipticals appear perfectly spherical, *E4* galaxies are somewhat football shaped, and *E7* ellipticals appear quite flattened, almost like the stereotypical UFO. Since ellipticals consist almost entirely of old stars, there is a noticeable absence of any nebulosity to them.

If a galaxy has no distinctive shape, then it is known as an *irregular* galaxy (Figure 6.11). The two closest irregular galaxies to the Milky Way are the Magellanic Clouds (although some authorities list the Large Magellanic Cloud as a loosely wound barred spiral). Neither is visible from the continental United States.

Cataloging the Universe

Just as we collect and catalogue everyday items according to size, type, color, and so forth, astronomers catalogue objects in the universe. Numerous listings of deep sky objects have been amassed over the course of human history.

The first organized attempt to catalogue the universe was made long before the invention of the telescope. In 150 B.C., the Greek

Figure 6.11
M82 (NGC 3034), in Ursa Major, is the sky's
most famous example of an irregular galaxy.
Photograph by Jack Newton.

scholar Hipparchus first recorded two strange cloudy objects in the sky that seemed to move with the stars. Today we know that Hipparchus saw the Beehive open cluster in Cancer and the Double Cluster in Perseus.

When he compiled his monumental *Almagest,* the second-century A.D. astronomer Ptolemy mentioned both the Double Cluster and the Beehive, as well as many other "nebulae." He also noted the Coma Berenices star cluster, a misty patch near the tail of Scorpius now known as open cluster "M7," and several stellar asterisms.

Other pretelescopic catalogues include those by Al Sufi (A.D. 964), Ulugh Begh (1437), and Tycho Brahe (1601). The most complete naked-eye survey of the night sky was made by Bayer in 1603 and became part of his famed *Uranometria* star atlas.

A whole new universe greeted the telescope-aided astronomer. Among the more extensive of the early telescopic catalogues were those by Hodierna (1653), Kirch (1682), Hevelius (1690), Halley (1715), De Cheseaux (1745–46), and Lacaille (1752). Even after the telescope came into being (and well into the 20th century), most deep sky objects were listed under the broad heading of "nebulae." Assigning this term to an object was actually more of a reflection on the optical instrument used (or lack thereof) and the then-current knowledge of the universe, rather than it was on the object's true nature. While some indeed proved to be interstellar clouds, most were eventually resolved into star clusters and galaxies.

In 1758, a French astronomer and comet hunter named Charles Messier began to compile what has become the most famous deep sky object catalogue of all. His involvement in the project was quite by accident. While searching for a comet, Messier came upon a hazy object near Zeta Tauri, which marks the tip of Taurus' easternmost horn. He thought for certain that he had discovered a new comet. Night after night, Messier returned to watch the curious apparition for any telltale motion against the background stars. Nothing happened. Finally, he was forced to conclude that what he had seen was not a comet, but rather something beyond the Solar System.

Noting its position, Messier began to compile a list of nebulous patches discovered by him and his peers. Today we know his initial entry as *M1*, the Crab Nebula. His notes read:

> what caused me to undertake the catalogue was the nebula I discovered above the southern horn of Taurus on September 12, 1758. . . . This nebula had such a resemblance to a comet, in its form and brightness, that I endeavored to find others, so that other astronomers would not confuse these nebulae with comets just beginning to shine.

Ironically, all of the comets that Messier discovered have faded into oblivion, but his catalogue of star clusters, nebulae, and galaxies remains to challenge amateur astronomers to this day.

Sir William Herschel, in England, compiled his own catalogue of over 2,500 star clusters and nebulae at the end of the 18th century. Some years later, his list was expanded by his son, John Herschel, who traveled to the Cape of Good Hope to study the southern sky. Pooling the two sets of observations, the younger Herschel published the *General Catalogue of Nebulae* in 1864.

Unlike the Messier catalogue, which simply lists the objects at random, the Herschels organized their tally according to each entry's visual appearance. The following eight categories cover all of the objects in their listing:

Class I.   Bright nebulae
Class II.   Faint nebulae
Class III. Very faint nebulae
Class IV.   Planetary nebulae
Class V.   Very large nebulae
Class VI.   Very compressed star clusters
Class VII. Compressed star clusters
Class VIII. Scattered star clusters

For example, the pair of clusters in Perseus' Double Cluster is listed in the *General Catalogue* as H-33-VI and H-34-VI. This translates into the pair's being the 33rd and 34th objects in the Herschels' list of Class VI, "very compressed star clusters." Such a classification system was subjective at best and confusing in many cases.

The *General Catalogue* remained the authoritative source on deep sky objects for only a short period of time. In 1888, John L. E. Dreyer published his *New General Catalogue of Nebulae and Clusters,* known as the NGC. Over 7,000 objects are included in the NGC, accounting for nearly all of the sky's brighter clusters, nebulae, and galaxies. Most of the Messier objects are also cross-referenced in the NGC. For instance, M1 is identified as NGC 1952 in Dreyer's listing.

To alleviate some of the confusion generated by the Herschels' complex classification system, Dreyer chose to catalogue the universe in sequence according to each object's right ascension coordinate. Beginning at right ascension 00 hours, the catalogue numbers increase moving eastward.

Subsequent discoveries were included in two *Index Catalogues* (IC). The first was published in 1895 and included objects found during the years 1889 to 1894, while the second was completed in 1908 and covered discoveries made from 1895 to 1907.

Many other catalogues have since been compiled, but nearly all concentrate on one specific type of object. In 1927, Edward Emerson Barnard compiled "A Catalogue of 349 Dark Objects in the Sky," published in Part I of *A Photographic Atlas of Selected Regions of the Milky Way.* Specific entries are referred to by their "B" numbers. For example, B33 identifies the famous Horsehead Nebula in Orion.

Most open clusters included in the Messier and NGC/IC listings are relatively small in apparent diameter. As a result, many large clusters were omitted from these classic references. Less well-known catalogues, such as those by Melotte (Mel) and Collinder (Cr), include many of these missed stellar aggregations. Among other works mentioned throughout Chapter 7 are those by Trumpler (Tr), Stock, and Harvard (H).

Anyone who has observed the night sky through binoculars has undoubtedly come upon interesting shapes and figures among the stars. While not actually physical clusters of stars in close proximity, these asterisms are nonetheless attractive to observe, frequently more so than some true clusters. Scattered throughout Chapter 7 are a dozen of these hitherto unlisted objects. This is my "Harrington" catalogue, abbreviated "Hrr." Rest assured that I am not trying to compete with the likes of Messier or the Herschels. In fact I cannot even claim to have discovered these objects; most were passed along by friends and fellow amateurs. I just wanted to take this opportunity to point out a few of the many interesting albeit unknown targets across the sky. "My" catalogue is anything but exhaustive. Why not add a few of your own?

Double stars have several catalogues all their own. Among the most popular are lists by the 19th century father-son team of Wilhelm and Otto Struve. Members of Wilhelm's list are signified in the next chapter by "Σ" while those from Otto's are identified by either OΣ or OΣΣ. Other double star inventories that remain popular today are those by Burnham and Dunlop. Perhaps the most comprehensive catalogue, entitled *A New General Catalogue of Double Stars* and abbreviated "ADS," was compiled in 1932 by R. G. Aitken. The ADS includes most of the double stars visible from northern skies. Each entry lists the pair's component magnitudes, separation, and position angle.

Variable stars are grouped together by their "home" constellation using capital Roman letters, beginning with "R," in the order in which they were discovered. For example, R Coronae Borealis was the first variable found in Corona Borealis, while S Scuti was the second discovered in Scutum, and so on. Once the ninth variable, denoted by "Z," was catalogued in a particular constellation, the next was identified by "RR," followed by "RS" to "RZ," then "SS" to "SZ," and continuing in this fashion until "ZZ" was discovered. After this, "AA" to "AZ" were assigned, then "BB" to "BZ," and so on. The system continues, excepting "JJ" to "JZ," until "QZ" is used, allowing 334 variable stars to be identified and labeled in a single constellation.

While this scheme would certainly seem adequate to cover all variables in a given constellation, many constellations, especially those along the main stream of the Milky Way, contain more than the system permits. In these cases, the 335th variable identified in a particular constellation is listed as "V335" (V for variable), with each subsequent discovery placed in numerical order.

In addition, a six-digit designation number is assigned to each variable. The designation number instantly tells an observer where the star is in the sky in epoch 1900.0 coordinates. For instance, R Coronae Borealis has been assigned 154428. The first four digits refer to its right ascension coordinate, namely, 15 hours 44 minutes. The last two numbers identify its declination, 28° north of the celestial equator. Should the variable lie south of the celestial equator, then the last two digits would either be underlined or have a minus sign placed in front of them. As an example, the bright variable R Hydrae is located at right ascension 13 hours 24 minutes, declination −22°. Therefore, its designation number may be written as either 1324<u>22</u> or 1324−22. Either way is correct, and both are understood to specify the same star.

Projects for the Amateur Astronomer

Searching for deep sky objects is enjoyed by an increasing number of amateurs and is a great way to hone your observational talents. Begin by looking for some of the brighter "showpiece" objects, and then slowly graduate to more difficult targets as your observational skills develop.

One of the most interesting and satisfying projects that an observer can engage in is finding as many Messier objects as possible. Of the 109 entries listed by Messier, over 80 are visible in 7×50 binoculars under suburban skies. Many are quite striking, while others are dim and barely perceptible. All will test your observational skills.

Begin your Messier project with the easier objects, listed in Table 6.1 under "Skill Level 1." All level-1 objects are visible through even the most modest binoculars and may even be seen with the naked eye under superior sky conditions.

Once the first tier has been conquered, move along to the "Skill Level 2" collection. These objects are also quite bright, but are located away from handy guide stars, so some hunting will be required. Again, under dark skies, some may be visible without any optical assistance.

"Skill Level 3" objects are seen only dimly in binoculars. These are easy to locate thanks to the presence of bright nearby guide stars, but are hard to find due to the objects' faintness.

Finally, "Skill Level 4" objects are seen as faint smudges of light and are positioned far from any easily found stars. Finding them is the ultimate test for an observer with binoculars and should only be attempted on the darkest, clearest nights.

A certificate is offered by the Astronomical League to those who find 70 Messier objects, to commemorate the triumph. (The Astronomical League is a national confederation of astronomy clubs across the United States.) To qualify for the award, you must either belong to the League through an affiliated club or be a member-at-large, and be able to produce a logbook noting which objects you have found. For the name and address of the closest affiliated club to you, contact the Astronomical League at 6235 Omie Circle, Pensacola, FL 32054.

Once a year, around the spring equinox, the Sun is positioned in the sky in such manner that all but one of the Messier objects can be observed in a single sunset-to-sunrise marathon. On that date, the Sun is in Pisces, and only M30, a globular cluster in Capricornus, is lost in the solar glare.

TABLE 6.1 Messier Objects Visible through Binoculars

Skill Level 1 No.	Constellation	Skill Level 2 No.	Constellation	Skill Level 3 No.	Constellation	Skill Level 4 No.	Constellation
4	Sco	2	Aqr	1	Tau	9	Oph
6	Sco	3	CVn	20	Sgr	14	Oph
7	Sco	5	Ser	28	Sgr	19	Oph
8	Sgr	10	Oph	29	Cyg	26	Sct
13	Her	11	Sct	32	And	30	Cap
15	Peg	12	Oph	40	UMa	33	Tri
16	Ser	23	Sgr	51	CVn	49	Vir
17	Sgr	27	Vul	54	Sgr	52	Cas
18	Sgr	34	Per	56	Lyr	53	Com
21	Sgr	39	Cyg	57	Lyr	58	Vir
22	Sgr	44	Cnc	65	Leo	59	Vir
24	Sgr	46	Pup	66	Leo	60	Vir
25	Sgr	47	Pup	69	Sgr	61	Vir
31	And	48	Hya	71	Sge	62	Oph
35	Gem	50	Mon	77	Cet	63	CVn
36	Aur	55	Sgr	80	Sco	64	Com
37	Aur	67	Cnc	101	UMa	68	Hya
38	Aur	78	Ori	103	Cas	70	Sgr
41	CMa	92	Her	110	And	72	Aqr
42	Ori	93	Pup			73	Aqr
43	Ori					75	Sgr
45	Tau					79	Lep
						81	UMa
						82	UMa
						83	Hya
						84	Vir
						85	Com
						86	Vir
						87	Vir

TABLE 6.1 (continued)

| Skill Level 1 | | Skill Level 2 | | Skill Level 3 | | Skill Level 4 | |
No.	Constellation	No.	Constellation	No.	Constellation	No.	Constellation
						88	Com
						89	Vir
						90	Vir
						94	CVn
						95	Leo
						96	Leo
						99	Com
						100	Com
						104	Vir
						105	Leo
						106	CVn
						107	Oph

The most hectic times during the marathon occur right after sunset and immediately before sunrise. The evening "rush hour" objects set shortly after the Sun and therefore must be seen as soon as the sky is dark enough. These include M31, M32, and M110 in Andromeda; M33 in Triangulum; and M79 in Lepus. M74, in Pisces, and M77, in Cetus, are also part of the early evening family, but they are too faint for detection in binoculars. (M77 is visible in binoculars in a dark sky but not when it is so close to evening twilight.) The morning "rush hour" begins around 3 A.M. and includes six members: M15 in Pegasus; M55 and M75 in Sagittarius; and M2, M72, and M73 in Aquarius. All are visible at night through binoculars, but M72, M73, and M75 may be lost in the light of the growing morning dawn.

By planning well in advance, anyone can run the Messier marathon. Pick a night close to the vernal equinox when the Moon is absent and the sky is clear and dark. For more details on the art of Messier marathoning, see the January 1985 issue of *Sky and Telescope.*

Star clusters, nebulae, and galaxies all have one thing in common—they require clear, dark, moonless skies to be seen best. Urban and suburban amateurs may find that local conditions do not readily permit their detection. Does this mean that deep sky observing is restricted to rural astronomers only? Not at all, because two groups of objects may be viewed even through light and air pollution: double stars and variable stars.

Many enjoyable hours may be spent in pursuit of double stars. They challenge both the quality of your binoculars' optics and the acuity of your eyesight. Test yourself against the binary stars described throughout the next chapter. Perhaps you will differ with the comments listed. You may be able to resolve a binary that is mentioned as unresolvable in binoculars, or you may see color where none is listed. No two people perceive the same object in the same way.

As observational skills develop, subtleties in celestial bodies become more and more obvious. Coloration of double stars is a good example of this. Many binaries will appear pure white when first viewed, but may slowly transform into colorful pairings as experience grows. A trick to help accent stars' colors is to defocus their images slightly.

Try observing binaries under a variety of conditions. Like the brighter planets, binaries are usually best seen through slightly hazy skies, when the atmosphere is steadiest. You will undoubtedly find that some of the stars listed are unresolvable under crystal clear, albeit turbulent, sky conditions, but are easy to resolve on hazy, steady nights.

Amateur observations of variable stars constitute a very important source of data contributing toward our understanding of the universe. The American Association of Variable Star Observers (AAVSO) is an international organization devoted to the study of these stars. Through systematic monitoring, observers keep a careful watch over variable star activity and report magnitude estimates on a monthly basis to AAVSO headquarters in Cambridge, Massachusetts. The AAVSO acts as liaison between amateur astronomers contributing data and professional astronomers requesting data. Monthly reports are combined to produce light curves of the stars' activity over the period.

Visually estimating the magnitude of a variable star is not as difficult as it may sound. The AAVSO has a wide variety of detailed charts available for selected variables. Each chart shows the targeted star and its surrounding neighborhood.

Figure 6.12 shows finder charts for four of the more interesting variable stars visible in binoculars. Each star is described under its respective constellation heading in Chapter 7. As an example of how to use a finder chart, consider the chart for Mira (Omicron Ceti), a classic long period variable. Notice that many of the nearby stars have numbers next to them. Each number represents the apparent

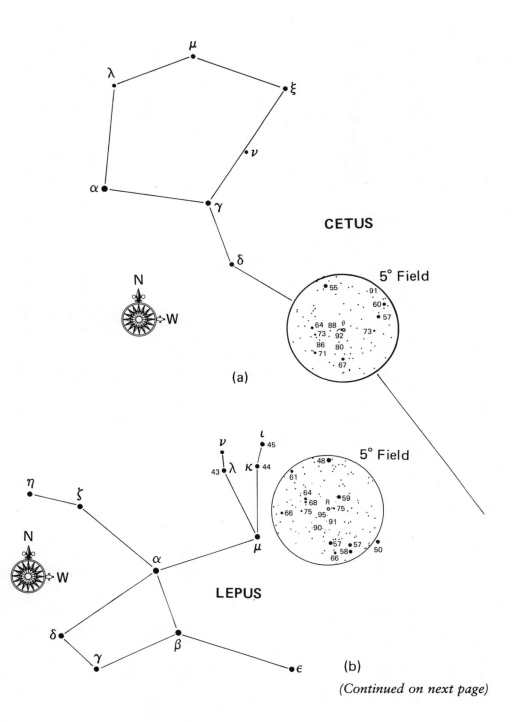

Figure 6.12
(a) The long period variable Mira (Omicron Ceti) is an ideal star for autumn observers.
(b) R Leporis, a long period star, is one of winter's finest variables for binocular study.
(c) The long period variable R Leonis is easy to spot in the spring sky. (d) Two fine
summer variables, Beta Lyrae and R Lyrae. The former is an eclipsing binary, while the
latter is a bright semi-regular sun.

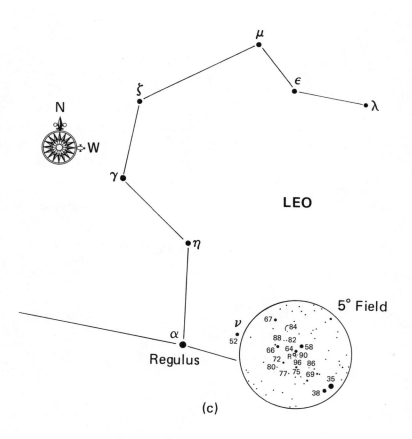

(c)

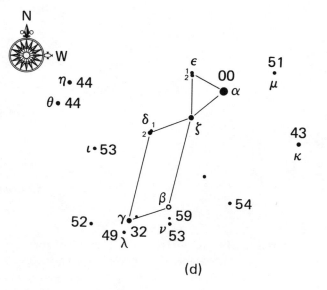

(d)

Figure 6.12 (continued)

magnitude, to the nearest tenth, of the indicated star. For example, the star marked "55" is magnitude 5.5, and so on.

Amateurs can successfully estimate the magnitude of a variable to an accuracy of 0.1 by comparing the variable to neighboring stars of known brightness. I like to call this process visual interpolation. Suppose that when you find Mira, it appears brighter than the nearby magnitude 7.3 (labeled 73) star, but fainter than the magnitude 5.7 (57) star. Then, check it against the magnitude 6.7 (67) comparison star to the south and perhaps with the magnitude 6.0 star to the variable's west. If Mira appears to be halfway between the two in brightness, then your "guesstimate" for its magnitude will be either 6.3 or 6.4. Now, double-check it against the 64 star to the east. If it seems closer to the 60 star than to the 64 star, then revise your estimate to perhaps 6.1 or 6.2. In essence, by comparing the variable to two stars of known, fixed brightness, one brighter and one fainter, an accurate reading may be made.

To note the precise time that an estimate of magnitude is made, variable star observers use the *Julian* calendar. Unlike the Gregorian calendar that we use in our everyday lives, the Julian calendar has no weeks, no months, and no years. It is simply the continual counting of days since its inception on January 1, 4713 B.C. One important difference between our day-to-day Gregorian calendar and the Julian calendar is that, while we begin our day at midnight, the Julian calendar starts its day at 1200 (noon) Universal Time.

If you are interested in pursuing the observation of variable stars, contact the AAVSO at 25 Birch Street, Cambridge, MA 02138. This fine organization will gladly send you information, sample finder charts, and Julian calendars.

Scanning the skies for new novae is a pastime pursued by relatively few amateur astronomers, but one which binoculars lend themselves to nicely. Like comet hunting, searching for novae requires great patience and determination on the part of the observer. Countless hours may be spent in vain searching for a never-before-seen exploding star. But to the victor go the spoils, for the discoverer of a nova is instantly catapulted to celebrity status in the astronomical community.

There is no way of predicting exactly where the next nova will flare; however, statistics show that most occur within 10° of the plane of the Milky Way. Moreover, the frequency of novae has always been greatest in the span between Cygnus and Sagittarius.

It would hardly seem practical to scan the entire galactic plane for novae. Indeed, unless it is unusually bright, a nova may be passed over on scanning too large an area. Instead, an observer should concentrate on relatively small patches of sky, where at least most of the visible stars may be memorized. Check the region carefully and note any distinctive star patterns and asterisms. This way, if an invader should suddenly burst forth in the future, its presence will be easily detected.

Just because a new point of light appears overnight in a star field does not necessarily mean it is a nova. Perhaps the interloper is a variable star on its way toward maximum brightness. To rule out the existence of a variable, check a detailed star atlas against the position of the suspected nova.

Of course, the "new star" might be an asteroid casually passing through the field of view. To double check against this possibility, consult either the *Minor Planet Bulletin* (see Chapter 4) or one of the astronomical periodicals or annuals listed in the appendix for the locations of asteroids visible at the time of sighting. Also, check to make sure that you have not "discovered" Uranus or Neptune!

If the object still remains a mystery, review recent *IAU Circulars* for announcements of any newly detected novae. The *Circulars* are available through subscription (see Chapter 4) or at many larger university and public libraries. If you cannot get hold of recent copies, call *Sky and Telescope's* "Skyline" recorded telephone message at 617-497-4168. Updated at least once a week, the recording highlights new discoveries in astronomy.

If, after checking all of these sources diligently, the mystery object's identity remains uncertain, then it is time to alert the authorities. Just as with a newly discovered comet, all reports of new novae should be addressed to the Center for Astronomical Telegrams at the Smithsonian Astrophysical Observatory in Cambridge, Massachusetts. The SAO may be reached through its Western Union telex number, 710-320-6842 (answerback ASTROGRAM CAM). In the message, the following must be included:

1. Your name, address, and telephone number.
2. The date and time (in U.T.) of the observation, in decimals of a day.
3. The right ascension and declination of the suspected nova.
4. The magnitude of the suspected nova.

Nova hunting does not require expensive equipment—just lots of perseverance and a little luck. Perhaps you will follow in the footsteps of British amateur astronomer G. E. D. Alcock. Alcock began his career as a nova hunter back in 1953. Armed only with a pair of binoculars and great determination, he has since discovered four novae. This feat makes him the most successful amateur nova hunter of the 20th century (although Peter Collins is hot on his heels, with three findings).

7

A Survey of the Night Sky

Each season, the evening sky offers something different for us to marvel at through binoculars. During the summer and winter months, the night side of the Earth is aimed toward the plane of our galaxy. The gentle rifts of the Milky Way stretch across the sky, providing spectacular views of rich star clouds and subtle nebulae.

Our nighttime window is turned away from the Milky Way in spring and autumn. During those seasons, we are looking outward toward the farthest reaches of the known universe. Distant galaxies, each a megalopolis of billions of suns, are sprinkled across the starry vault. While most are too faint to be seen through binoculars, there are several striking examples of each galactic type to view. Some are easily found, while others will test our skills as observers. When we look toward the galaxies, our gaze is traveling across millions of light years.

Seven-power glasses are all that are needed to see most of the objects listed in this guide, while giant glasses are required to see the "really faint stuff" (lower than 9th magnitude). Each target offers a unique view; those of special interest are listed in boldface print and described in detail.

For reasons of space, abbreviations for various types of deep sky objects are employed throughout the data lists in this chapter. Table 7.1 lists these abbreviations and their meanings.

TABLE 7.1 Abbreviations for Types of Deep Sky Objects

Abbreviation	Translation	Abbreviation	Translation
*	Star	DN	Diffuse nebula
* *	Double or multiple star	Dk	Dark nebula
Vr	Variable star	PN	Planetary nebula
OC	Open (galactic) cluster	Gx	Galaxy
GC	Globular cluster		

Several deep sky object catalogues and lists are referred to throughout this "constellation dissection" chapter. While you may be familiar with many of them, some will probably be new. Table 7.2 may be used to translate the catalogues.

Each constellation table also lists each entry's type, right ascension and declination coordinates for epoch 2000, visual magnitude (unless the value is followed by a "p," for photographic magnitude), apparent size, separation or period of variability as appropriate, and any pertinent comments.

Assigning magnitude to nonstellar objects is tricky, and it can be even trickier to interpret. Just because an object is listed as relatively bright does not necessarily mean that it is easy to spot. Visibility is strongly dependent on many factors, including magnitude, apparent size, and atmospheric conditions. For instance, a 7th-magnitude galaxy with an apparent diameter of, say, 5' of arc may actually appear brighter than a 6th-magnitude galaxy with a 10' span. This discrepancy is caused by the effective surface brightnesses of the galaxies. Remember that the magnitudes listed are the equivalent brightnesses *if each target could be shrunk to a stellar point.* Expanding each to its measured diameter causes them to fade away, in some cases quite rapidly.

Another important consideration is just how the value of the magnitude was obtained. If you look up the same object in several independent references, odds are that few will list the exact same magnitude for the object. Some will list the object's visual magnitude, while others will quote its photographic or blue magnitude. In general (though not universally), an object's photographic magnitude will be about a full magnitude fainter than its visual magnitude. This is especially important to remember when searching for variable stars at or near their extreme values of magnitude.

The size of an extended object is also a topic of hot debate, especially in regard to how it was measured. Not surprisingly, most deep sky objects will appear significantly larger in long-exposure photographs taken at professional observatories than they will

TABLE 7.2 Deep Sky Object Cross-Reference Listing

Object Type	Designation in Chapter	Author and/or Catalogue
**	ADS	Aitken, *New General Catalogue of Double Stars*
Dk	B	Barnard, *Catalogue of 349 Dark Objects in the Sky*
**	B	van den Bos
OC	Basel	Basel
Dk	Be	Bernes
OC	Berk	Berkeley
OC	Biur	Biurakan
OC	Blanco	Blanco
OC	Bochum	Bochum
**	BrsO	Brisbane Observatory, Australia
**	β	Burnham (S.W.)
OC	Cr	Collinder
OC	Do	Dolidze
OC	DoDz	Dolidze-Dzimselejsvili
**	Δ	Dunlop
OC	Fein	Feinstein
**	h	Herschel (John)
**	H	Herschel (William)
OC	Haffner	Haffner
OC	Harvard	Harvard
*	HD	Draper, *Henry Draper Catalogue*
OC	Hogg	Hogg
OC,Dk	Hrr	Harrington
OC,DN	IC	Dreyer, *Index Catalogue*
OC	Isk	Iskudarian
Dk	LDN	Lynds
OC	Lynga	Lynga
OC,GC,DN, PN,G	M	Messier
OC	Mel	Melotte
OC	Mrk	Markarian
OC,GC,DN, PN,G	NGC	Dreyer, *New General Catalogue of Nebulae and Clusters of Stars*
**	OΣ	Struve (Otto), *Pulkovo Catalogue*

(Continued on next page)

TABLE 7.2 (continued)

Object Type	Designation in Chapter	Author and/or Catalogue
**	OΣΣ	Struve (Otto), *Pulkovo Catalogue Supplement*
OC	Pismis	Pismis
PN	PK	Perek and Kohoutek, *Catalogue of Galactic Planetary Nebulae*
OC	Roslund	Roslund
OC	Ru	Ruprecht
**	S	South
**	SHJ	South and Herschel (John)
Dk	SL	Sandqvist and Lindroos
OC	St	Stock
OC	Steph	Stephenson
**	Σ	Struve (F.G.W.), *Dorpat Catalogue*
OC	Tombaugh	Tombaugh
OC	Tr	Trumpler
OC	Upgren	Upgren
**	Wnc	Winnecki

through binoculars. Keep this in mind when you are searching for some of the objects listed. You may stumble upon a nebula with a listed diameter of perhaps 30′, but through binoculars it appears less than half that size. "Is Harrington crazy?" you ask yourself. Not necessarily, as the values quoted here are based on photographic measurements. In other words, don't always blame me!

This same column is the place to find the period of variability for all referenced variable stars and the angular separation between double stars.

Where appropriate, I have tried to include data in the "Comments" column that will enhance the reader's knowledge and enjoyment of the objects observed. For instance, the comments following all double-star entries list, in order, the stars' position angle (P.A.), the year the position angle was measured, the stars' number in Aitken's *New General Catalogue of Double Stars* (ADS), and other pertinent or interesting information. Comments on variable stars include each star's type or class, as well as any popular proper names and information for finding the star.

General remarks on the various nonstellar deep sky objects include their Messier catalogue entry (if appropriate), common nicknames, hints for finding the object, and other little tidbits. Open clusters are followed by the term "asterism" if the association between members of the cluster is not physical, while galaxies are always accompanied by their Hubble classification.

Each highlighted object impresses me as something special. Accordingly, rather than just blandly describe the appearance of each, I have tried to make the objects come alive. If you disagree with my choices, keep in mind that much of observational astronomy is (dare I say) subjective and that beauty is, after all, in the eye of the beholder.

I must confess one fault up front. While I have gone to great lengths to make this a "global" book, I remain a citizen of the northern hemisphere. Throughout the descriptions to come, I will frequently refer to a constellation's or an object's visibility according to season. These are northern hemisphere seasons. Keep in mind that the seasons below Earth's equator are the reverse of those above it. I hope southern hemisphere readers will forgive me!

There is a wealth of deep sky objects awaiting your scrutiny. Enjoy the view as you tour the universe through your binoculars, but in doing so, do yourself one favor. Perhaps as a sign of the times, people always seem to be in a rush. No sooner do they finish one task, and they are on to the next, never allowing themselves to bask in the moment. As you move from one target to the next, pause for a bit. Don't just look at the universe . . . *see* the universe. Absorb the beauty that the heavens hold for everyone.

Andromeda (And)

Object	Typ	R.A. (2000) h m	Dec. (2000) ° '	Mag.	Size/Sep/ Period	Comments
R	Vr	00 24.0	+38 35	5.8–14.9	409.33 days	Long Period Variable
NGC 205	Gx	00 40.4	+41 41	8.0	17'	E6 M110 M31 companion
NGC 221	Gx	00 42.7	+40 52	8.2	8' × 6'	E2 M32 M31 companion
NGC 224	Gx	00 42.7	+41 16	3.5	160' × 40'	Sb M31 Andromeda Galaxy
56	**	01 56.2	+37 15	5.7,6.0	190"	300° (1928); 1534
NGC 752	OC	01 57.8	+37 41	5.7	50'	
NGC 891	Gx	02 22.6	+42 21	10.0	14' × 3'	Sb; Edge on
NGC 7662	PN	23 25.9	+42 33	8.9p	32" × 28"	
NGC 7686	OC	23 30.2	+49 08	5.6	15'	

Andromeda, the mythological daughter of King Cepheus and Queen Cassiopeia, is a prominent member of the northern hemisphere's "royal family" of autumn. Portrayed as a pair of arcing lines of stars, the Princess holds several deep sky objects of interest to observers with binoculars. It is especially famous as the home of our Milky Way's "sister galaxy."

M31 (NGC 224), the renowned Andromeda Galaxy (Figure 7.1), at 2.2 million light years away, is the closest major galaxy to our own. Seen by the naked eye as a dim smudge just northwest of Nu Andromedae, the Andromeda Galaxy has been spotted since at least the 10th century, when it was first noted by Persian astronomers. The German astronomer Simon Marius became the first person to view M31 telescopically in 1611, although it was not recognized as a separate galaxy for another 313 years.

Using binoculars under moderately dark skies, an observer may trace the glow of M31 for about 3°. A pronounced central nucleus highlights the galactic halo. Sharp-eyed observers peering through large glasses might even be able to spot one of two dark dust lanes that encircle the outer rim of M31.

M32 (NGC 221) appears as a tiny circular glow just south of M31's bright core. The brightest satellite galaxy of M31, M32 is a small E2 system that shines at 8th magnitude and spans 8' × 6'. While it may be spotted with 7× glasses, 10× or greater magnification is recommended for a better view.

Figure 7.1
The Andromeda Galaxy, M31 (NGC 224), framed by M32 (NGC 221)
below and M110 (NGC 205) above. Photograph by Johnny Horne using
an 8-inch f/1.5 Schmidt camera, hypered Technical Pan 2415 film, and a
five-minute exposure.

M110 (NGC 205) is a second satellite galaxy of M31. Although
catalogues list it as being slightly brighter than M32, M110 is far
more difficult to glimpse due to its low surface brightness. (For more
about surface brightness versus magnitude, see the discussion under
"M33" in Triangulum.) To spot M110, larger glasses are a must.
Look for a nondescript oval glow to the northwest of M31, at about
twice the distance as M32.

NGC 752 is a large, loosely structured open cluster. It may be found
near a small triangle of stars about 5° south-southwest of Gamma
Andromedae. Measuring about 50′ of arc across, NGC 752 holds
about 75 stars within its grasp, not all of which are resolvable through
binoculars.

NGC 7662 is a challenging planetary nebula that is bright enough to
be seen in 7× glasses, although its 30 arc-second disk will appear

stellar. Only its soft blue-green glow sets it apart from surrounding stars. The 13th-magnitude central star of NGC 7662 is well below the threshold of visibility of binoculars.

Antlia (Ant)

Object	Typ	R.A. (2000) h m	Dec. (2000) ° '	Mag.	Size/Sep/ Period	Comments
S	Vr	09 32.3	−28 38	6.4–6.9	0.648 days	Eclipsing Binary
T	Vr	09 33.8	−36 37	8.9–9.8	5.90 days	Cepheid
RR	Vr	09 35.7	−39 54	9.4–10.9p		Irregular
U	Vr	10 35.2	−39 34	8.1–9.7p		Irregular

Antlia, the Air Pump, lies in a star-poor region of the early spring sky. First portrayed in about 1752 by Lacaille, this constellation is rich in faint, telescopic galaxies, but holds little to attract the binocular enthusiast. Only four variable stars are within range of most glasses.

S Antliae is the easiest variable in Antlia to find and follow with binoculars. Although its variation is only half a magnitude, its period of just over 15 hours can allow an observer to follow the star from maximum to minimum in a single night.

T Antliae is a short-period variable star classified as a Cepheid. Its magnitude fluctuates from 8.9 to 9.8 over a 5.90-day cycle. Giant binoculars should be able to follow it over its entire range.

Apus (Aps)

Object	Typ	R.A. (2000) h m	Dec. (2000) ° '	Mag.	Size/Sep/ Period	Comments
Theta	Vr	14 05.3	−76 48	6.4–8.6p	119 days	Semi-Regular
VZ	Vr	16 16.3	−74 02	8.2–17.5p	385 days	Long Period Variable
Delta[1+2]	**	16 20.3	−78 42	4.7,5.1	103″	12°(1918)
WW	Vr	16 31.7	−74 59	9.0–16.8p	267 days	Long Period Variable

Apus, the Bird of Paradise, lies far to the south in the late-spring sky and is invisible from midnorthern latitudes. Though not well stocked with bright deep sky objects, it does present a spectacular double star and a few interesting variables.

Theta Apodis is a semi-regular variable star. Over a period that averages 119 days, it may be seen to fluctuate from about visual magnitude 5 to almost 8. (Recall our discussion of visual versus photographic magnitude at the beginning of the chapter.) The reddish tint of this spectral type-M sun stands out in striking contrast against the starry surroundings.

Delta^{1+2} **Apodis** consists of a wide pair of 5th-magnitude stars. Even the smallest opera glasses will unveil the ruddy color of this striking stellar duo. The two stars are separated by 103″ of arc and form a small triangle with unrelated Gamma Apodis, a few degrees to the east.

Aquarius (Aqr)

Object	Typ	R.A. (2000) h m	Dec. (2000) ° ′	Mag.	Size/Sep/ Period	Comments
V	Vr	20 46.8	+02 26	7.6–9.4	244.0 days	Semi-Regular
T	Vr	20 49.9	−05 09	7.2–14.2	202.10 days	Long Period Variable
NGC 6981	GC	20 53.5	−12 32	9.4	6′	M72
NGC 6994	OC	20 58.9	−12 38	9.0	3′	M73; asterism
NGC 7009	PN	21 04.2	−11 22	8.4	26″	Saturn Nebula
NGC 7089	GC	21 33.5	−00 49	6.5	13′	M2
NGC 7293	PN	22 29.6	−20 48	6.5	900″ × 720″	Helix Nebula
S	Vr	22 57.1	−20 21	7.6–15.0	279.27 days	Long Period Variable
R	Vr	23 43.8	−15 17	5.8–12.4	386.96 days	Z And type

Aquarius, the Water Bearer, holds the 11th spot along the zodiac. Representing a man pouring water from a jar, this star group traces its origins to early Babylonian times. While it affords little for the naked eye observer, Aquarius offers many interesting targets for those with binoculars.

M72 (NGC 6981) is the faintest of the 29 globular clusters listed in the Messier catalogue. Found about 3° south of Mu Aquarii, M72 was discovered in August 1780 by Pierre Méchain. Rated 9th magnitude, it is detectable as a slightly fuzzy point of light in 7× glasses. Giant binoculars may make it a bit more obvious, but do little else to improve the visual impression of this 60,000-light-year-distant swarm.

M73 (NGC 6994) is erroneously included as an open cluster in both the Messier and NGC catalogues. Now known to be merely a small asterism of four faint stars about 1½° east of M72, it was recorded by Messier as a cluster with nebulosity. This cloudlike effect is only an illusion due to low power. Higher magnified telescopic views and photographs fail to show any nebulous wisps.

NGC 7009 lies about 3° farther east of M73. Known as the Saturn Nebula because of its faint, almost ringlike extensions, NGC 7009 is visible in binoculars as a greenish point of light and is famous for having one of the highest surface brightnesses of any planetary. This is due to unusually strong ultraviolet radiation emitted from its central star, which is too faint to be detected through binoculars. The greenish cast is caused by double-ionized oxygen.

M2 (NGC 7089) is the brightest nonstellar object in Aquarius and is a true showpiece of the autumn sky. This easily seen globular cluster was first glimpsed in 1746 by Maraldi and is visible through nearly all binoculars as a not-quite-stellar "star." Look for its 13′-diameter disk about 9° west of Alpha Aquarii and 6° north of Beta Aquarii. Large glasses will show that M2 is not perfectly round, but instead has a slightly oval disk. This effect is probably due to the cluster's rapid rotation. Estimated to lie about 37,000 light years away, M2 contains no fewer than 100,000 stars.

NGC 7293 (Figure 7.2), known popularly as the Helix Nebula, is unlike any other planetary nebula in the sky. While most appear as little more than starlike points, the Helix measures nearly 1/4° across! Under prime conditions, this celestial smoke ring looks like a round, hazy patch of dim gray light. Giant binoculars hint at the cloud's subtle texture, but the central star will remain unconfirmed. Due to its low surface brightness and large expanse, NGC 7293 is frequently much easier to pick out in the wider fields of binoculars than with comparatively narrow-field telescopes. When we look at the Helix, our gaze is traveling an estimated 450 light years, making NGC 7293 the closest planetary nebula to Earth.

Figure 7.2
The Helix Nebula (NGC 7293) in Aquarius, as photographed by Martin
Germano. Exposure time was 45 minutes through a 135-mm lens
stopped down to f/4.8 on hypered Technical Pan 2415 film.

Aquila (Aql)

Object	Typ	R.A. (2000) h m	Dec. (2000) ° '	Mag.	Size/Sep/ Period	Comments
NGC 6709	OC	18 51.5	+10 21	6.7	13'	
NGC 6738	OC	19 01.4	+11 36	8.3p	15'	
B132	Dk	19 04.1	−04 28		16' × 8'	40' NW of Lambda Aql
V	Vr	19 04.4	−05 41	6.6–8.4	353 days	Semi-Regular
15	**	19 05.0	−04 02	5.5,7.2	38"	209°(1959);12007
B133	Dk	19 06.1	−06 50		10' × 3'	2° S of Lambda Aql
R	Vr	19 06.4	+08 14	5.5–12.0	284.2 days	Long Period Variable
B135–6	Dk	19 07.4	−03 55		50' × 30'	
NGC 6755	OC	19 07.8	+04 14	7.5	15'	
OΣΣ 178	**	19 15.3	+15 05	5.7,7.8	90"	268° (1925)
B137–8	Dk	19 15.6	+00 13		180' × 10'	
B142–3	Dk	19 40.7	+10 57		80' × 50'	3° NW of Altair

Aquila, the Eagle, may be found flying high along the mainstream of
the Milky Way on warm summer nights. Highlighted by the brilliant

stellar lighthouse Altair, this region is a beautiful one to just sit back and slowly scan with binoculars. Several noteworthy objects, as well as many interesting, though unnamed, asterisms and star patterns, are found within Aquila.

NGC 6709 is a collection of about two dozen stars set within a 13′ area. Unfortunately, most of this open cluster's suns are too faint for detection by binoculars. Instead, their dim light blurs together to form a ghostly haze against the background sky. NGC 6709 may require some searching before it will be captured, but remember, the mark of a successful observer is patience and persistence.

NGC 6738 may be described in a similar fashion to NGC 6709. Once located, it reveals only a few of its brighter stars, which are about 9th magnitude. Larger glasses should have little trouble yielding a glimpse of these stars, and even 7× binoculars show a few of them set within the glow cast by fainter luminaries.

Barnard 133 is a small dark nebula found about 2° south of Lambda Aquilae. Look for a "hole" in the sky about one-third the size of the Full Moon, and that will be Barnard 133. Searching for dark nebulae should only be attempted on the clearest nights, far from all interfering sources of light pollution.

OΣΣ 178 is a double star found near Zeta Aquilae in far northern Aquila. Even the smallest field glasses should have no trouble resolving the inconspicuous pair, once found. Nearly 90″ of arc separate these faint yellow and white suns.

Barnard 142 and **Barnard 143** collectively form a fairly conspicuous dark patch located 1½° west of Gamma Aquilae. Nicknamed the "Fish on a Platter" Nebula, it is seen in larger binoculars as a 30′-diameter cloud with two "horns" extending further west.

Ara (Ara)

Object	Typ	R.A. (2000) h m	Dec. (2000) ° ′	Mag.	Size/Sep/ Period	Comments
R	Vr	16 39.7	−57 00	6.0–6.9	4.425 days	Eclipsing Binary
NGC 6188	DN	16 40.5	−48 47		20′ × 12′	
NGC 6193	OC	16 41.3	−48 46	5.2	15′	
NGC 6200	OC	16 44.2	−47 29	7.4	12′	
NGC 6208	OC	16 49.5	−53 49	7.2	16′	

Ara (continued)

Object	Typ	R.A. (2000) h m	Dec. (2000) ° ′	Mag.	Size/Sep/ Period	Comments
NGC 6204	OC	16 46.5	−47 01	8.2	5′	
NGC 6250	OC	16 58.0	−45 48	5.9	8′	
Harvard 13	OC	17 05.4	−48 11		15′	
IC 4651	OC	17 24.7	−49 57	6.9	12′	
NGC 6362	GC	17 31.9	−67 03	8.3	11′	
NGC 6397	GC	17 40.7	−53 40	5.6	26′	
U	Vr	17 53.6	−51 41	7.7–14.1	225.21 days	Long Period Variable

Ara, the Altar, offers little to the naked eye, since its brightest stars are only 3rd magnitude. Nonetheless, it is a splendid area for surveying with binoculars.

NGC 6193 is a rich open cluster of 30 stars, although binoculars may not show that many. The brightest cluster member is the quadruple star **h4876,** first catalogued by John Herschel. The system's primary sun is a 6th-magnitude landmark to watch for when searching for the cluster. Unfortunately, its companion stars are either too faint or too close to it to be detectable in binoculars.

To the southwest of NGC 6193 is a large region of bright and dark interstellar clouds spanning more than 3°. The central portion of this nebula is catalogued as **NGC 6188.** It will prove extremely difficult to spot visually, either through binoculars or a telescope, even under optimum conditions.

NGC 6208 is an attractive little open cluster found north of Zeta Arae. Here, observers will find a small 7th-magnitude glow representing about 60 cluster stars. The brightest of these shine at 9th magnitude, making them dimly perceptible through most binoculars.

NGC 6250 rides the Ara-Scorpius border. This is a poor harvest of about 15 stars, with the brightest being about 8th magnitude. Large binoculars should be used to positively identify this weak 8′ diameter open cluster. A few individual points of light can be seen against the glow of unresolved stellar members. While the cluster itself may not prove too impressive, the surrounding region is stunning. There is great beauty to be found in this area just by casually sweeping back and forth across these overflowing star clouds of the Milky Way.

IC 4651 may be found just west of 3rd-magnitude Alpha Arae. Observers should find this open cluster a fairly easy catch. Its 70 stars combine to 7th magnitude and are concentrated across 12′ of arc. However, only the largest glasses are able to resolve any of them individually, as the brightest are only 11th magnitude.

NGC 6397 is believed to be the nearest globular cluster to the Solar System. Lying only 8,200 light years away, it presents an attractive face through binoculars. Although no separate stars are visible, a definite "grainy" surface texture is seen through large glasses, as if stellar resolution were imminent. If it were visible from more northerly latitudes, NGC 6397 would be one of the better known globulars. Due to its far southern declination, however, it is destined to remain a little-observed beauty.

Aries (Ari)

Object	Typ	R.A. (2000) h m	Dec. (2000) ° ′	Mag.	Size/Sep/ Period	Comments
Lambda	**	01 57.9	+23 36	4.9,7.7	37″	46° (1933); 1563; 9 Ari
R	Vr	02 16.1	+25 03	7.4–13.7	186.78 days	Long Period Variable
30	**	02 37.0	+24 39	6.6,7.4	39″	274° (1937); 1982; colorful
T	Vr	02 48.3	+17 31	7.5–11.3	316.6 days	Semi-Regular
U	Vr	03 11.0	+14 48	7.2–15.2	371.13 days	Long Period Variable

Aries, the Ram, is a small, rather desolate autumn constellation that is best known for being the first sign of the zodiac. The naked eye sees the Ram as a lone 2nd-magnitude star and two 3rd-magnitude suns set in a narrow triangle. Binoculars do only a little better. Most reveal a trio of variables, as well as a pair of doubles, among the faint stars of Aries.

Lambda Arietis is the easier of the two double stars to glimpse. Its 4.9- and 7.7-magnitude component suns are separated by about 37″ of arc and shine pure white. Binoculars of less than 7× will find this duo a challenge to resolve, but more powerful glasses should have no trouble splitting them.

30 Arietis is an equally challenging binary star system. Here, the 7.4-magnitude companion lies about 39″ of arc away from the 6.6-magnitude primary sun. The pair has been described as yellow and pale lilac by some observers, although others see no color to either star. What do you see?

Auriga (Aur)

Object	Typ	R.A. (2000) h m	Dec. (2000) ° '	Mag.	Size/Sep/ Period	Comments
NGC 1664	OC	04 51.1	+43 42	7.5	18'	
AB	Vr	04 55.8	+30 33	6.9–8.4		Irregular
NGC 1778	OC	05 08.1	+37 03	7.7	7'	
R	Vr	05 17.3	+53 35	6.7–13.9	457.51 days	Long Period Variable
Hrr 4	OC	05 19	+33		75'	**Asterism**
NGC 1857	OC	05 20.2	+39 21	7.0	6'	
UV	Vr	05 21.8	+32 31	7.4–10.6	394.42 days	Long Period Variable
Cr 62	OC	05 22.5	+41 00	4.2p	28'	
NGC 1893	OC	05 22.7	+33 24	7.5	12'	
NGC 1907	OC	05 28.0	+35 19	8.2	7'	
NGC 1912	OC	05 28.7	+35 50	6.4	21'	M38
NGC 1960	OC	05 36.1	+34 08	6.0	12'	M36
Stock 10	OC	05 39.0	+37 56		25'	
NGC 2099	OC	05 52.4	+32 33	5.6	24'	M37
RT	Vr	06 28.6	+30 30	5.0–5.8	3.728 days	Cepheid
UU	Vr	06 36.5	+38 27	7.8–10.0	234 days	Semi-Regular
NGC 2281	OC	06 49.3	+41 04	5.4	15'	

I always look forward to the appearance of **Auriga,** the Charioteer, as it rises above the treetops on late autumn nights, for it signals the arrival of the magnificent winter sky. Auriga's brightest jewel, Capella (Alpha Aurigae), is the first of the brilliant "Winter 8" circle of stars to dawn from midnorthern latitudes.

Auriga brings with it many splendid deep sky objects for the binocularist's enjoyment. Home to three of the finest open clusters in the Messier catalogue, the Charioteer also holds many lustrous clusters from the NGC and other lesser known catalogues.

Harrington 4, consisting of **16, 17, 18, 19,** and **IQ Aurigae,** is the first stop in our tour of Auriga. The naked eye can easily pick up the stars' combined glow as an oblong haze near the constellation's center. Although it is not a true open cluster, binoculars yield a very pleasant, star-rich view, especially through wide-angle glasses. In all, about 15 stars brighter than 9th magnitude fill the 75' span of this asterism.

Collinder 62 is a little-observed group found about 5° south of Capella. With an apparent diameter equal to that of the Full Moon, this open cluster's overall brightness rating of 4th magnitude is a bit misleading: although most of its stars are 8th magnitude or fainter, a single 5th-magnitude sun boosts the group's brightness to the listed

Figure 7.3
A wide view of the constellation Auriga featuring three fine winter open clusters: M38 (NGC 1912), M36 (NGC 1960), and M37 (NGC 2099), right to left, respectively. Right of center is the asterism referred to here as Harrington 4. Photograph by Bernard Volz using a 135-mm f/2.8 lens, ISO 1600 film, and a 17-minute exposure.

value. Still, the cluster is evident in glasses and is certainly worth a visit.

M38 (NGC 1912), discovered in 1749 by LeGentil, is the western-most of three Messier objects shown in Figure 7.3 and offers a pleasant sight in binoculars. Most observers will see it as a circular glow with a few faint stars scattered across. Many have commented that, through giant glasses, the stars of M38 appear to be arranged in the pattern of the Greek letter "pi." Others say that they resemble a cross. Do you see a pattern in the stars of M38, and if so, what does it look like?

M36 (NGC 1960) is about half the size of M38 and contains about five dozen 8th-magnitude and fainter stars. Of these, only about ten may be seen through 7× glasses. The remaining suns blur into a gentle stellar fog. On exceptionally clear nights, M36 takes on an almost three-dimensional effect against the background Milky Way field.

Stock 10 is another one of Auriga's many unsung open clusters. Though rather sparsely populated, this stellar aggregation stands out surprisingly well, as over half of its 15 stars are 8th magnitude or brighter. The cluster's two brightest stars form a relatively close 6th-magnitude binary system, which helps to further clarify the cluster's identity. Stock 10 can be found about 4° west-northwest of Theta Aurigae and 4° north of M36.

M37 (NGC 2099) is the most striking open cluster in Auriga and a joy to behold in nearly all instruments. Since its discovery in 1764 by Messier, many observers have commented on its beauty. In his book *Celestial Objects for Common Telescopes,* the Reverend T.W. Webb noted, "it is extremely beautiful, one of the finest in its class," while others have likened it to a sprinkling of stardust. About 150 stars belong to M37 and are swarmed together in a tight collection. A few individual points of light may be resolved in low-power glasses, while more powerful giant binoculars will show many stars strewn across the glimmer of still fainter, unresolved suns.

Boötes (Boo)

Object	Typ	R.A. (2000) h m	Dec. (2000) ° '	Mag.	Size/Sep/ Period	Comments
S 656	**	13 50.4	+21 17	6.8,7.3	86″	208° (1923)
NGC 5466	GC	14 05.5	+28 32	9.1	11′	
Iota	**	14 16.2	+51 22	4.9,7.5	39″	33° (1942); 9198
RX	Vr	14 24.2	+25 42	8.6–11.3p	340 days	Semi-Regular
R	Vr	14 37.2	+26 44	6.2–13.1	223.40 days	Long Period Variable
RV	Vr	14 39.3	+32 32	7.9–9.9p	137 days	Semi-Regular
RW	Vr	14 41.2	+31 34	8.0–9.5p	209 days	Semi-Regular
Delta	**	15 15.5	+33 19	3.5,8.7	105″	79° (1976); 9559
Mu	**	15 24.5	+37 23	4.3,6.5	108″	171° (1956); 9626

The arrival of **Boötes,** the Herdsman, in the eastern evening sky is a welcome sight to many in the northern hemisphere, for it signals that spring has arrived. Twentieth-century imaginations will find it easier to trace out a kite or an ice cream cone in this region, rather than a human form, as our ancestors did.

Alpha Boötis, better known as Arcturus, is the brightest star north of the celestial equator. Shining at a dazzling magnitude −0.06, its distinctive orangish hue is unmistakable through a telescope, binoculars, or the naked eye. It is found only 37 light years away, making Arcturus one of our Sun's more impressive neighbors.

NGC 5466, found near the border shared with Canes Venatici, is a challenging globular cluster for the observer with giant binoculars. With the cluster shining at only 9th magnitude, its detection is strongly dependent on sky conditions, so wait for that special night before searching it out. Even then, it appears as only a dim smudge of gray light just to the west of a 7th-magnitude star.

Iota Boötis, a double star of particular interest to binocular users, is found in the extreme northwest corner of the constellation. Separated by 39″ of arc, the magnitude 7.5 secondary sun is separable from the 5th-magnitude primary with 7× binoculars on nights of steady seeing. Note the unrelated 6th-magnitude star to the pair's east. It may mistakenly give the impression that Iota is actually a triple star system.

Mu Boötis is another double star worthy of note. Discovered by F. G. W. Struve in 1835, Mu is easily seen as a binary in all glasses. Its magnitude 6.5 secondary star lies 108″ of arc away from the magnitude 4.3 primary. Both exhibit little color.

Caelum (Cae)

Object	Typ	R.A. (2000) h m	Dec. (2000) ° ′	Mag.	Size/Sep/ Period	Comments
R	Vr	04 40.5	−38 14	6.7–13.7	390.95 days	Long Period Variable

Caelum, the Chisel, is a small constellation wedged between Eridanus and Horologium to the west and Columba to the east. Named by the 17th-century astronomer Lacaille, Caelum skims the southern horizon from the northern hemisphere's more temperate regions, but it attracts little attention, as its brightest stars are only 4th magnitude.

R Caeli resides about three-fourths of the way from Alpha to Beta Caeli. Shining with a distinct reddish glint, this long-period variable fluctuates between magnitudes 6.7 and 13.7 over a 391-day cycle.

Camelopardalis (Cam)

Object	Typ	R.A. (2000) h m	Dec. (2000) ° '	Mag.	Size/Sep/ Period	Comments
Stock 23	OC	03 16.3	+60 02		15'	
OΣΣ 36	**	03 40.0	+63 52	6.8,8.6	46"	69° (1923); 2650
Tombaugh 5	OC	03 47.8	+59 03	8.4	17'	
S 436	**	03 49.3	+57 07	6.5,7.3	58"	75° (1975)
Hrr 3	OC	04 00	+63			Asterism
NGC 1502	OC	04 07.7	+62 20	5.7	8'	
11	**	05 06.1	+58 58	5.4,6.5	180"	8° (1924)
Cr 464	OC	05 22	+73	4.2	120'	
Σ 1051	**	07 26.6	+73 05	7.1,7.8	31"	82° (1935); 6028
NGC 2403	Gx	07 36.9	+65 36	8.4	17' × 10'	Sc
OΣΣ 90	**	08 02.5	+63 05	6.0,8.4	49"	82° (1924)
NGC 2655	Gx	08 55.6	+78 13	10.1	5' × 4'	SBa

Of all the northern circumpolar constellations, **Camelopardalis** is certainly the least distinct. It was created by Jakob Bartsch in 1624 to represent a camel, but was later reidentified as a giraffe. With none of its stars shining brighter than 4th magnitude, amateurs tend to shy away from the wide variety of deep sky objects found within. However, by using binoculars as extensions of our eyes, we may star-hop among the Giraffe's faint stars to search out its treasures.

Stock 23 resides on the Camelopardalis-Cassiopeia border, not far from the plane of the Milky Way. Of the 25 stars that comprise this little-known open cluster, half a dozen shine between 7th and 9th magnitude and are visible in 7× glasses. Look for this small, rectangular stellar knot about 10° north-northwest of Alpha Persei.

Harrington 3 is a long string of stars measuring about 2° in extent and found about 30° due east of Epsilon Cassiopeiae. Two dozen stars ranging from 5th to 9th magnitude lie along its breadth. Toward the southeastern end of this asterism is NGC 1502 (see next) and UV Camelopardalis, a semiregular variable star that flickers between magnitudes 7.5 and 8.1 with an average period of 294 days.

NGC 1502 is a bright, easy-to-see but hard-to-locate open cluster adjacent to Harrington 3 in central Camelopardalis. Shimmering at 6th magnitude, it may be glimpsed on crystalline evenings as a hazy glow about 15° east of Cassiopeia's "W." Binoculars reveal its richness as a misty circular patch of light sprinkled with four 8th-magnitude suns.

Collinder 464 is a large, loosely packed open cluster found about 17° away from Polaris and 8° northeast of Alpha Camelopardalis. Although its stellar density is quite low, this attractive cluster stands out surprisingly well against its star-poor surroundings. It appears distinctly rectangular in shape, with approximately 50 magnitude 5 and fainter suns scattered across a 2° × 1° area. Few amateurs are aware that the cluster exists. Take the time on the next clear evening to discover it.

NGC 2403 is one of the brightest non-Messier galaxies found north of the celestial equator. Located about 5° northwest of Omicron Ursae Majoris (the Bear's "nose"), it is revealed as a large, oval glow highlighted with a brighter central core set among a wide rectangular asterism. Studies indicate that NGC 2403 resides 8 million light years away and is one of the closest major galaxies beyond the Local Group.

Cancer (Cnc)

Object	Typ	R.A. (2000) h m	Dec. (2000) ° '	Mag.	Size/Sep/ Period	Comments
R	Vr	08 16.6	+11 44	6.1–11.8	361.60 days	Long Period Variable
β 584	**	08 39.9	+19 33	6.9,7.2	45",93"	156°, 241° (1952); 6915; in M44
NGC 2632	OC	08 40.1	+19 59	3.1	95'	M44, Beehive or Praesepe
Iota	**	08 46.7	+28 46	4.2,6.6	31"	307° (1968); 6988
NGC 2682	OC	08 50.4	+11 49	6.9	30'	M67
X	Vr	08 55.4	+17 14	5.6–7.5	195 days	Semi-Regular
RT	Vr	08 58.3	+10 51	7.1–8.6	60 days	Semi-Regular
RS	Vr	09 10.6	+30 58	6.2–7.7p	120 days	Semi-Regular

Cancer, the Crab, is a faint, unimpressive spring constellation nestled between the brighter stars of Gemini and Leo. Consisting of 4th- and 5th-magnitude stars, it is easily overlooked by casual stargazers, especially through brighter urban and suburban skies. Yet buried within are a few of the season's most notable binocular objects.

R Cancri is one of four notable variable stars in Cancer that are suitable for study through binoculars. Found about 3° north of Beta

Cancri, R is a classic long period variable that may be followed through binoculars over nearly its entire cycle. Peaks in brightness, averaging magnitude 6.1, arrive every 362 days, with minima of about magnitude 11.8 occurring in between.

M44 (NGC 2632), popularly known as the Praesepe or Beehive Cluster, is a huge open cluster that is visible to the unaided eye on clear, dark nights. Recorded as early as the third century B.C. as a mysterious cloudy spot, this magnificent swarm of stars was still classified as a nebula in Bayer's *Uranometria* star atlas of 1603. The true nature of the Praesepe came to light just seven years later, when Galileo viewed it through his first crude telescope. Today, even the smallest field glasses will resolve many of its member stars spread across 95' of arc. Over 200 stars are counted as belonging to M44, about 75 of which are brighter than 10th magnitude.

One of the most interesting sights to look for within M44 is the triple star **Burnham 584** (β584). The three almost equally bright points of light may be found forming a small triangle just south of the cluster's center. Separated by 45" and 93", respectively, all three stars present little trouble for 7× glasses.

Iota Cancri is a challenging binary star for most binoculars. Its yellowish magnitude 4.2 primary star is accompanied by a bluish magnitude 6.6 companion located about 31" away. Seven-power is probably the lowest magnification that can separate the stars, while higher power will reveal Iota to be one of the most striking binary stars in the northern spring sky.

M67 (NGC 2682) is often overlooked in favor of the more spectacular Beehive cluster. This is unfortunate, for M67 is a pleasing open cluster in its own right. With their combined glow equivalent to 7th magnitude, over 500 stars scattered across 30' of arc compose M67. The brightest of these stars just reach 10th magnitude. Most glasses, therefore, will see only the misty glow of stars too faint to resolve, but giant binoculars should also display a few points of light buried within. Astronomers now feel that this 2,500-light-year-distant collection is one of the oldest open clusters in the Milky Way, perhaps second only to the ancient group NGC 188 in Cepheus. The age of M67 is estimated to be about 10 billion years.

Canes Venatici (CVn)

Object	Typ	R.A. (2000) h m	Dec. (2000) ° '	Mag.	Size/Sep/ Period	Comments
NGC 4258	Gx	12 19.0	+47 18	8.3	18' × 8'	Sb+p M106
NGC 4449	Gx	12 28.2	+44 06	9.4	5' × 4'	Ir+
Upgren 1	OC	12 35.0	+36 18		15'	
NGC 4631	Gx	12 42.1	+32 32	9.3	15' × 3'	Sc
Y	Vr	12 45.1	+45 26	7.4–10.0p	157 days	Semi-Regular
NGC 4736	Gx	12 50.9	+41 07	8.2	11' × 9'	Sb-p M94
17	**	13 10.1	+38 30	6.0,6.2	84"	297° (1922); 8805
NGC 5005	Gx	13 10.9	+37 03	9.8	5' × 3'	Sb-
NGC 5055	Gx	13 15.8	+42 02	8.6	12' × 7'	Sb+ M63
V	Vr	13 19.5	+45 32	6.5–8.6	191.9 days	Semi-Regular
NGC 5194	Gx	13 29.9	+47 12	8.4	11' × 8'	Sc M51; **Whirlpool Galaxy**
NGC 5195	Gx	13 30.0	+47 16	9.6	5' × 4'	**P M51 companion**
NGC 5272	GC	13 42.2	+28 23	6.4	16'	**M3**

Canes Venatici, the Hunting Dogs, is another difficult-to-discern constellation of the northern spring sky. Lying just southeast of Ursa Major, it is home to many challenging deep sky objects for the enthusiast with binoculars.

M106 (NGC 4258) is one of dozens of distant galaxies that call Canes Venatici home. Readily found about 7° to the northwest of Beta Canum Venaticorum, M106 appears as a relatively large elliptical glow less than a degree west of a 6th-magnitude foreground star. The galaxy's nucleus appears decidedly nonstellar with higher magnifications. In long-exposure photographs, M106 spans 18' × 8', but it appears about half that size when viewed through binoculars.

Upgren 1 is a bright, coarse open cluster known to few amateurs. Have you ever seen it? The odds are you probably haven't, but I'll bet you have been all around it if you have ever explored Canes Venatici. It is easily visible in binoculars, so you might have even bumped into it without realizing it was a cluster. Only ten stars scattered across 15' of arc and found about 5° southwest of Cor Caroli (Alpha Canum Venaticorum) belong to the group.

Seven-power glasses reveal four magnitude 7 stars (three of which are strung in a short, straight line), along with a couple of fainter suns. Larger binoculars add a few more dim points of light to this 450 light-year-distant group. Although studies show it to be a true cluster, Upgren 1 never really impresses me as much more than a chance asterism. What do you think?

M94 (NGC 4736) is a tightly wound Sa spiral located northeast of the halfway point between Alpha and Beta Canum Venaticorum. Look for a nearly circular halo surrounding a fairly bright starlike center. Though rated at 8th magnitude like M106 to its west, M94 appears noticeably brighter to most observers. This effect is due to the comparative sizes of the two galaxies. M94 exhibits a higher surface brightness than its neighbor, since its light is spread across a smaller $11' \times 9'$ area. These dimensions translate to a real diameter of 64,000 light years at the galaxy's distance of 20 million light years away.

M63 (NGC 5055) is found almost halfway between the stars Alkaid (Eta Ursae Majoris, the end star in the handle of the Big Dipper) and Cor Caroli (Alpha Canum Venaticorum). Discovered in 1779 by Pierre Méchain, this 9th-magnitude Sb spiral looks like an ill-defined smudge of light set close to an 8th-magnitude star. Giant glasses reveal it as distinctly cigar shaped and absent of any bright central nucleus.

M51 (NGC 5194), the Whirlpool Galaxy, is a textbook example of a spiral galaxy (Figure 7.4). Lord Rosse was the first to detect its pinwheel-like structure in 1845 while viewing through his mammoth 72-inch reflector from Parsonstown, Ireland. Initially, astronomers thought that such "spiral nebulae" were actually solar systems in formation. It was not until the 1920s, when Edwin Hubble conceived the true galactic organization of the universe, that they were recognized as remote galaxies.

Most binoculars readily show M51. It dwells among the 6th- and 7th-magnitude stars of a small stellar trapezoid a few degrees southwest of Alkaid (Eta Ursae Majoris). Look for a round 8th-magnitude glow punctuated by a conspicuous stellar nucleus.

NGC 5195 is a small irregular satellite galaxy of M51. Long-exposure photographs reveal that the two galaxies are physically bound to one another by a bridge of stars and nebulosity. Unfortunately, NGC 5195 glows weakly at nearly 10th magnitude and is not likely to be seen in anything less than $11 \times 80's$. It may require even larger glasses to be seen. Look for a subtle protrusion on the northern edge of M51.

M3 (NGC 5272) is one of the brightest globular clusters in the entire sky. Discovered in 1764 by Charles Messier, it is seen through binocu-

Figure 7.4
M51 (NGC 5194) and NGC 5195 in Canes Venatici. Photograph taken by Johnny Horne using a 12½-inch Newtonian reflector, hypered Technical Pan 2415 film, and a 40-minute exposure.

lars as a round, 6th-magnitude nebulous "star" some 16′ in diameter. Although individual stars in the cluster require at least a four-inch telescope to be seen, a "grainy" appearance may be noted in giant binoculars, as if some of the stars are very close to resolution. Modern estimates show that over half a million stars populate M3, making it one of the largest members of the Milky Way's family of globular clusters. It is thought to lie about 40,000 light years away and measure 200 light years in diameter.

Canis Major (CMa)

Object	Typ	R.A. (2000) h m	Dec. (2000) ° ′	Mag.	Size/Sep/ Period	Comments
NGC 2287	OC	06 47.0	−20 44	4.6	38′	M41
Cr 121	OC	06 54.2	−24 38	2.6	50′	Omicron CMa Cluster
W	Vr	07 08.1	−11 55	6.4–7.9		Irregular
NGC 2345	OC	07 08.3	−13 10	7.7	12′	
NGC 2360	OC	07 17.8	−15 37	7.2	13′	

CMa (continued)

Object	Typ	R.A. (2000) h m	Dec. (2000) ° '	Mag.	Size/Sep/ Period	Comments
NGC 2362	OC	07 18.8	−24 57	4.1	8'	**Tau CMa Cluster**
R	Vr	07 19.5	−16 24	5.7–6.3	1.136 days	Eclipsing Binary
NGC 2367	OC	07 20.1	−21 56	7.9	3.5'	
Ru 16	OC	07 23.2	−19 27		11'	
Cr 140	OC	07 23.9	−32 12	3.5	42'	
NGC 2374	OC	07 24.0	−13 16	8.0	19'	
Δ 47	**	07 24.7	−31 49	5.5,7.6	99"	342° (1922)
NGC 2383	OC	07 24.8	−20 56	8.4	6'	
NGC 2384	OC	07 25.1	−21 02	7.4	2.5'	

Standing obediently by Orion's side, **Canis Major,** the Large Dog, appears ready to help its master do battle with Taurus the Bull in our winter sky. For amateurs, the only battle to be waged within Canis Major is deciding which deep sky object among its wide array to look at first.

Signaling the arrival of Canis Major in our winter sky is brilliant Sirius (Alpha Canis Majoris), nicknamed the Dog Star. Its name may come from the Greek word *seirios,* which means "scorching." Other sources claim the term has Egyptian or Celtic roots.

Sirius dazzles us at apparent magnitude −1.4, the brightest of any star in our night sky. However, it shines at absolute magnitude +1.5, which is far from the intrinsically brightest star known. Classified as a hot type-A white star, Sirius is a little over twice the mass of the Sun and lies only 8.7 light years away. A small white dwarf known as Sirius B, or the "Pup," orbits Sirius A once every half-century. Seeing Sirius B is one of the greatest challenges to astronomers with telescopes.

M41 (NGC 2287), shown in Figure 7.5, was first recorded by Aristotle in 325 B.C. as a "cloudy spot" about 4° south of Sirius. Today's observers can easily duplicate Aristotle's discovery, as M41 stands out clearly without optical aid if viewed far from city lights.

M41 is one of the most satisfying clusters visible through binoculars, as even 7× glasses are able to pick out up to two dozen of its stars. All sparkle like sapphires against black velvet. Many appear as double or multiple stellar combinations. When fully resolved, M41 is found to hold about 80 stars within its gravitational grip.

Collinder 121 is found 4° farther south of M41. While its northern neighbor is easy to identify, this 2,300-light-year-distant horde is

Figure 7.5
M41 (NGC 2287) in Canis Major. Photograph taken by George Viscome using a 500-mm lens, ISO 400 film, and a six-minute exposure.

difficult to discern because of its starry locale. Aiding us in seeking it out is 4th-magnitude Omicron Canis Majoris, which dwells on the group's northern border. To this star's south is a pair of 6th-magnitude suns and about a dozen 7th-magnitude and fainter stars, which account for nearly all 20 components of the cluster.

NGC 2362 is a small, tightly packed open cluster easily pinpointed around Tau Canis Majoris. Unfortunately, through low-power binoculars, the light from this beacon tends to overwhelm the 40 or so cluster stars. Higher magnification glasses resolve up to a dozen magnitude and fainter points.

From surveying the stars within NGC 2362, astronomers believe this to be one of the youngest known clusters. However, the studies are inconclusive when it comes to Tau Canis Majoris. Is it a true member of NGC 2362, or simply a chance foreground star? If Tau lies within the cluster, then it is one of the most intrinsically luminous stars known. At the cluster's estimated distance of 5,000 light years, Tau would shine more than 50,000 times brighter than the Sun!

Collinder 140 is a bright, but sparse cluster of stars found just inside Canis Major's extreme southeastern corner. About 30 stars ranging from 5th to fainter than 9th magnitude are set in a triangular pattern about ¾° across. The clustering effect is best seen in 6× to 9× glasses, as increased magnification only serves to scatter the group even more.

Canis Minor (CMi)

Object	Typ	R.A. (2000) h m	Dec. (2000) ° '	Mag.	Size/Sep/ Period	Comments
R	Vr	07 08.7	+10 01	7.3–11.6	337.38 days	Long Period Variable
Do 26	OC	07 30.1	+11 54		24'	In front of 6 CMi
S	Vr	07 32.7	+08 19	6.6–13.2	332.94 days	Long Period Variable

Set within the brilliant winter sky is the small, rather bland constellation **Canis Minor,** the Small Dog. Except for the brilliant star Procyon, Canis Minor holds little to draw observers away from the more spectacular constellations of the season. Only a pair of long period variables and a weak open cluster are visible within.

Dolidze 26 is found about 7° northwest of Procyon, near the border shared with Gemini. Binoculars disclose only two individual stars (the brighter being **6 Canis Minoris**), as well as the dimmest hint of the cluster's few, unresolvable stars.

Capricornus (Cap)

Object	Typ	R.A. (2000) h m	Dec. (2000) ° '	Mag.	Size/Sep/ Period	Comments
Alpha[1+2]	**	20 18.1	−12 33	3.6,4.2	378"	291° (1924); 13645; optical
Beta[1+2]	**	20 21.0	−14 47	3.4,6.2	205"	267° (1922)
NGC 7099	GC	21 40.4	−23 11	7.5	11'	M30

Capricornus is described as a "sea-goat," a curious mythological creature that has the head and torso of a goat and the tail of a fish. Marking the tenth constellation along the ecliptic, Capricornus is framed by a large, crooked triangle of many faint naked eye stars and holds few deep sky objects for binoculars. Only a pair of wide double stars and a lone globular cluster will be visible.

Alpha[1+2] Capricorni is a broad optical binary star that marks the northwest corner of Capricornus' triangle. Keen-sighted naked-eye observers can easily distinguish the two stars, separated by better than 6' of arc. Binoculars add a pale yellow tint to each of the suns. Both Alpha[1], at magnitude 3.6, and Alpha[2], at magnitude 4.2, are type-G stars, but are actually nowhere near each other in space. They just happen to lie along the same line of sight from Earth. Both are in fact true binary stars in their own right, although their companions are too close to the respective primaries to be seen through binoculars.

Beta[1+2] **Capricorni** is another bright, easily split binary star. Lying about 2½° southeast of Alpha Capricorni, Beta contains a magnitude 3.4 primary star paired with a magnitude 6.2 secondary. Observers with sharp color perception might detect a blue tint to Beta[2], whereas Beta[1] remains pure white. The stars' common proper motion indicates that they are a true physical pair with nearly one trillion miles between them. In addition, Beta[2] is itself a tight binary, bringing the total to three stars in this 150-light-year-distant system.

M30 (NGC 7099) lies in the southeastern part of Capricornus, less than a degree west of 5th magnitude 41 Capricorni. Messier was first to find this globular cluster, in August 1764. He described it as simply "a nebula." In 1783, William Herschel became the first person to detect the true nature of M30. Through binoculars, we see pretty much what Messier saw—that is, a round, misty patch of light drawing to a brighter core.

Carina (Car)

Object	Typ	R.A. (2000) h m	Dec. (2000) ° '	Mag.	Size/Sep/ Period	Comments
NGC 2516	OC	07 58.3	−60 52	3.8	30'	
Δ 74	**	08 57.0	−59 14	5.1,6.8	40"	75° (1917)
NGC 2808	GC	09 12.0	−64 52	6.3	14'	
NGC 2867	PN	09 21.4	−58 19	9.7p	11"	
IW	Vr	09 26.9	−63 38	7.9–9.6p	67.5 days	RV Tauri type
R	Vr	09 32.2	−62 47	3.9–10.5	308.71 days	Long Period Variable
NGC 3114	OC	10 02.7	−60 07	4.2	35'	
S	Vr	10 09.4	−61 33	4.5–9.9	149.49 days	Long Period Variable
NGC 3199	DN	10 17.1	−57 55		22'	
HR	Vr	10 22.9	−59 37	8.2–9.6p		Irregular; S Dor type
NGC 3247	OC	10 25.9	−57 56	7.6	7'	
IC 2581	OC	10 27.4	−57 38	4.3	8'	
YZ	Vr	10 28.3	−59 21	8.2–9.1	18.163 days	Cepheid
NGC 3293	OC	10 35.8	−58 14	4.7	40'	
Bochum 9	OC	10 35.8	−60 08	6.3		
NGC 3324	OC/ DN	10 37.3	−58 38	6.7	6'	
Mel 101	OC	10 42.1	−65 06	8.0	14'	

Car (continued)

Object	Typ	R.A. (2000) h m	Dec. (2000) ° ′	Mag.	Size/Sep/ Period	Comments
Bochum 10	OC	10 42.2	−59 09	6.2		
Cr 228	OC	10 43.0	−60 01	4.4	15′	On NGC 3372
IC 2602	OC	10 43.2	−64 24	1.9	50′	Theta Car cluster
NGC 3372	DN	10 43.8	−59 52	5.0	120′	Eta Carinae Nebula
Tr 14	OC	10 43.9	−59 34	5.5	5′	On NGC 3372
Δ 99	**	10 44.3	−70 52	6.3,6.5	63″	75° (1917)
VY	Vr	10 44.6	−57 34	6.9–8.1	18.990 days	Cepheid
Tr 15	OC	10 44.8	−59 22	7.0	3′	On NGC 3372
Tr 16	OC	10 45.1	−59 43	5.0	10′	Eta Carina Nebula cluster
Bochum 11	OC	10 47.3	−60 06	7.9		On NGC 3372
IX	Vr	10 50.4	−59 59	9.0–10.0p	400 days	Semi-Regular
BZ	Vr	10 54.1	−62 03	8.9–10.8p	97 days	Semi-Regular
AG	Vr	10 56.2	−60 27	7.1–9.0p		Irregular; S Dor type
Cr 236	OC	10 57.0	−61 02	7.7p	8′	
U	Vr	10 57.8	−59 44	5.7–7.0	38.768 days	Cepheid
NGC 3496	OC	10 59.8	−60 20	8.2	9′	
XZ	Vr	11 04.2	−60 59	8.1–9.1	16.650 days	Cepheid
Fein 1	OC	11 06.0	−59 49	4.7		
NGC 3532	OC	11 06.4	−58 40	3.0	55′	
NGC 3572	OC	11 10.4	−60 14	6.6	20′	
Hogg 10	OC	11 10.7	−60 22	6.9	3′	
Cr 240	OC	11 11.2	−60 17	3.9	25′	
Tr 18	OC	11 11.4	−60 40	6.9	12′	
NGC 3590	OC	11 12.9	−60 47	8.2	4′	
Stock 13	OC	11 13.1	−58 55	7.0	3′	
IC 2714	OC	11 17.9	−62 42	8p	12′	

Carina is the southernmost of the four modern constellations formed from the archaic star group Argo Navis (the others are Puppis, Pyxis, and Vela). Carina is only partially seen from North America and Europe. Its brightest star is brilliant Canopus (Alpha Carinae), an FO giant and the second brightest star in the night sky at magnitude −0.72. Though normally considered a far southern target, Canopus can be seen from as far north as +30° latitude given a good southerly view. All of Carina is a celestial bonanza for amateur astronomers, with some of the finest deep sky objects located within its borders. Many enjoyable hours may be spent in search of its riches.

NGC 2516 is a striking open cluster in western Carina. Spanning ½° across, it is visible to the naked eye as a hazy patch between Alpha Pictoris and Epsilon Carinae. Fully one-third of its 100 stellar components are resolvable in 7 × 50 binoculars, while giant glasses offer a dazzling view of the star-filled field. With the exception of a lone ruddy sun toward the cluster's center, its stars shine pure white. NGC 2516 spans 15 light years and is 1,200 light years from Earth.

NGC 2808 is a noteworthy globular cluster located about 4° west of Nu Carinae. Its many 13th- to 15th-magnitude stars are too faint to be seen individually through binoculars, but their combined brightness yields a 6th-magnitude object that can be glimpsed with the unaided eye on clear nights. Through most glasses, NGC 2808 looks like a celestial ball of fluff highlighted with a bright center.

NGC 3114 is another absolutely stunning open cluster. Sadly, it only skims the horizon from northern latitudes, but for those fortunate enough to be at latitude +20° or below, it is a memorable sight. Up to three dozen stars ranging from magnitude 6 to magnitude 10 are visible through 7× binoculars, with the number growing as the size of the instrument increases. One view of NGC 3114 and you will surely add it to your "favorite clusters" list, as I have.

IC 2581 is a much tighter open cluster than either NGC 2516 or NGC 3114. Easily found surrounding 5th-magnitude P Carinae, most of its three dozen stellar citizens require 15× or greater to be resolved. Smaller instruments show only the brightest four or five members set within a small nebulous glow.

NGC 3293 is another rich, tightly packed open cluster that is striking in low- and high-power binoculars alike. About 50 stars from magnitude 6 to magnitude 13 belong to the group, but most observers will spot only about the brightest ten or so. Surrounding NGC 3293 is a large, faint cloud of nebulosity. Apparent in photographs of the area, this cloud probably cannot be sighted through anything less than large amateur telescopes.

NGC 3324 marks a fairly bright and extensive region of emission nebulosity in northern Carina. Seen as an irregular misty haze, NGC 3324 is excited into fluorescence by the tight double star h 4338, one of the faintest of the half-dozen stars superimposed on the cloud.

IC 2602 is one of the brightest open clusters around, yet it seems doomed to obscurity due to its far southern position in the sky. Visible well only from latitude +10° and southward, this fine cluster contains orangish 3rd-magnitude **Theta Carinae** surrounded by about 60 other stars ranging from 4th to less than 9th magnitude and spread across

Figure 7.6
The Eta Carinae Nebula (NGC 3372), one of the showpieces of the far southern sky. Photograph by Jack Newton.

nearly a full degree. At about 490 light years away, IC 2602 is one of the closest open clusters to us.

A sometimes unnoticed faint elliptical blur just to the south of IC 2602 is **Melotte 101.** Nearly 15 times farther away from Earth than its neighbor, Melotte 101 contains about 50 dim stars, with none brighter than 10th magnitude.

NGC 3372 (Figure 7.6) is the finest example of a diffuse nebula found anywhere in our skies. The amazing Eta Carinae Nebula, NGC 3372, is a monstrous glowing cloud seemingly blossoming forth like a huge ghostly orchard. Measuring two full degrees across, the nebula's many dark rifts divide it into several distinct regions, each with many stars embedded within. Even the smallest binoculars reveal some of the complexities of this marvelous object, and it grows into an indescribably beautiful sight through 11× glasses.

The larger, teardrop-shaped nebulous patch contains the open cluster **Trumpler 16** and the star **Eta Carinae** itself. Eta is a remarkable novalike star, first noted by Edmund Halley in 1677. At the time, Halley estimated its brightness at 4th magnitude. By 1730, it had increased to 2nd magnitude, but it fell back to 4th magnitude over the next half-century. Eta continued to fluctuate with a general upward trend until April 1843, when it reached magnitude −0.8 (slightly brighter than neighboring Canopus). Since that time, it has continued to vary, but has never achieved the same intense magnitude. Presently, Eta is about 6th magnitude. Its spectrum shows the star to be about ten solar masses. The presence of heavy elements in large amounts has led some astronomers to conclude that Eta Carinae is an old star and a leading candidate for the next Milky Way supernova!

Along with Eta Carinae, Trumpler 16 contains about ten stars tightly packed into the center of NGC 3372. No fewer than five additional open clusters—**Trumpler 14 and 15, Bochum 10 and 11, and Collinder 228**—are also seen superimposed onto the nebula. All appear as tight clumps of few stars.

Feinstein 1 is usually passed over by most observers. Found about 3° due east of Eta Carinae, it appears as a miniaturized version of Corona Borealis, with a half-circle of eight magnitude 7 stars and at least twice as many fainter suns scattered throughout. Feinstein 1 is a charming object, made even more attractive by the rich star field within which it is set.

NGC 3532 stands out as something extra special among the many superb open clusters within Carina. Herschel wrote that this was the most brilliant cluster he had ever seen, a sentiment echoed by many other observers as well. About 400 stars populate NGC 3532, with over 60 bright enough to be seen in 7 × 50 binoculars. An extensive study carried out in 1930 by Harlow Shapley determined that over 90% of the cluster members are brilliant, hot type-A stars. Current estimates place the group at 1,300 light years away and about 25 light years across.

NGC 3572 is a small but easily found galactic cluster. The brightest of its 35 suns are 7th magnitude, the faintest barely 14th. Binoculars show a small, faint nebulous glow of unresolved stars peppered with a few of the brighter stellar images.

Collinder 240 is a bright knot of five stars seen immediately to the southeast of NGC 3572. Binoculars readily resolve an additional six or seven stars within the cluster's 25′ extent. In all, about 30 stars belong to Collinder 240.

Trumpler 18 is located less than ½° south of the NGC 3572–Collinder 240 pair. The least interesting group of the three, it appears as little more than a nebulous wisp around four 9th-magnitude suns.

Cassiopeia (Cas)

Object	Typ	R.A. (2000) h m	Dec. (2000) ° '	Mag.	Size/Sep/ Period	Comments
TV	Vr	00 19.3	+59 08	7.2–8.2	1.813 days	Eclipsing Binary
T	Vr	00 23.2	+55 48	6.9–13.0	444.83 days	Long Period Variable
NGC 129	OC	00 29.9	**+60 14**	6.5	21'	
NGC 225	OC	00 43.4	+61 47	7.0	12'	
NGC 281	OC/ DN	00 52.8	+56 37	7.4p	23' × 27'	
Gamma	Vr	00 56.7	**+60 43**	1.6–3.0		Irregular; Gamma Cas prototype
NGC 457	OC	01 19.1	**+58 20**	6.4	13'	
NGC 581	OC	01 33.2	**+60 42**	7.4	6'	M103
NGC 654	OC	01 44.1	+61 53	6.5	5'	
NGC 659	OC	01 44.2	+60 42	7.9	5'	
NGC 663	OC	01 46.0	+61 15	7.1	16'	
Cr 463	OC	01 48.4	+71 57	5.7	36'	
Stock 5	OC	02 04.5	+64 26		15'	
Stock 2	OC	02 15.0	**+59 16**	4.4	60'	"Muscle Man" Cluster
OΣΣ 26	**	02 19.7	+60 02	6.9,7.4	63"	200° (1925)
Mrk 6	OC	02 29.6	+60 39	7.1	4.5'	
IC 1805	OC/ DN	02 32.7	+61 27	6.5	22'	
NGC 1027	OC	02 42.7	**+61 33**	6.7	20'	
RZ	Vr	02 48.9	+69 38	6.2–7.7	1.195 days	Eclipsing Binary
IC 1848	OC/ DN	02 51.2	+60 26	6.5	12'	
Cr 33	OC	02 59.3	+60 24	5.9p	40'	
Cr 34	OC	03 00.9	+60 25	6.8p	25'	
Tr 3	OC	03 11.8	+63 15	7.0p	23'	
V	Vr	23 11.7	+59 42	6.9–13.4	228.83 days	Long Period Variable
NGC 7635	DN	23 20.7	+61 12		15' × 8'	Bubble Nebula
Hrr 12	OC	23 20	**+62 30**		60'	**Asterism**
NGC 7654	OC	23 24.2	**+61 35**	6.9	13'	M52
Stock 12	OC	23 37.2	+52 26		20'	
Rho	Vr	23 54.4	+57 30	4.1–6.2	320 days	Semi-Regular
NGC 7789	OC	23 57.0	**+56 44**	6.7	16'	
R	Vr	23 58.4	+51 24	4.7–13.5	430.46 days	Long Period Variable
NGC 7790	OC	23 58.4	+61 13	8.5	17'	

With its distinctive "W" shape found high overhead during the autumn months, **Cassiopeia,** the Queen, is usually one of the first constellations learned by astronomers in mid-northern latitudes. Thanks to the Milky Way passing directly through, Cassiopeia is further glorified with an abundance of deep sky objects.

NGC 129 is a large, bright open cluster. Most binoculars reveal from half a dozen to a dozen suns nestled among the collective glow of numerous other stars too faint to resolve. About 35 luminaries belong to this 5,200-light-year distant swarm, including the Cepheid variable DL Cassiopeiae. DL fluctuates between magnitudes 8.6 and 9.3 over the course of eight days and is located near the cluster's center, just north of a pair of 9th-magnitude stars.

Gamma Cassiopeiae, the central star in Cassiopeia's "W," is the prototype for a class of irregular variable stars. Gamma Cassiopeiae stars are thought to be rapidly spinning spectral type-B stellar infernos that exhibit random magnitude fluctuations of up to about 1½ magnitudes. Gamma itself flickers between magnitudes 1.6 and 3.0.

NGC 457 resides just northwest of 5th-magnitude Phi Cassiopeiae and 7th-magnitude HD 7902. Easily resolvable into many points of light, this fine open cluster consists of two southward-arcing rows of stars along with a smattering of fainter suns. Although Phi and HD 7902 are unrelated foreground suns, their presence certainly adds to the cluster's elegance. The brightest star in the cluster itself shines at magnitude 8.6 and is clearly orangish in color.

M103 (NGC 581) is a fairly compact open cluster seen to the northeast of Delta Cassiopeiae. Pierre Méchain is credited with the discovery of the group in 1781. Consisting of about 25 magnitude 10 and fainter stars, M103 spans only 6' of arc and remains mostly unresolvable in less than 15× glasses. All binoculars, however, will reveal a string of four stars immediately to the cluster's southeast. While thought to be unrelated to M103, these suns greatly add to its scenic beauty.

Stock 2 is one of my favorite "unsung" deep sky objects. While many observers have no doubt seen it, few ever realize that they have, as this fine open cluster lies only 2° north of the famous Double Cluster in Perseus. While it cannot hope to compete with that magnificent sight, Stock 2 is still striking in nearly all glasses. Spanning a full degree across, the group consists of 50 stars shining at 8th magnitude or fainter. Seven-power binoculars unveil many cluster members sprayed onto the glow of fainter suns. Take a long look at the stars of Stock 2. As a friend mentioned to me not long ago, the brighter stars almost look like a headless stick man flexing his muscles. On the next clear autumn night, *before* you immediately go to the Double Cluster, pause first for a look at Stock 2. He's been there all along!

NGC 1027 is an attractive open cluster found about 4½° east-southeast of Epsilon Cassiopeiae. Through giant glasses, many of the group's brighter stars are plainly evident, while smaller binoculars lend the cluster a hazy appearance. In all, 40 magnitude 9 and fainter stars belong to this 3,300-light-year-distant starry flock.

M52 (NGC 7654) is one of the richest open clusters north of the celestial equator. Discovered by Messier in September 1774, M52 contains about 100 stars crammed into a relatively small 13′ area. Several individual suns are resolvable in binoculars, but the rest blur into a nebulous mass, which is exactly how Messier first described the group.

Just to the north of M52 is **Harrington 12,** a wide triangular asterism of about a dozen 5th- to 9th-magnitude stars. While not designated as a true open cluster, it is a very attractive group in low-power binoculars.

NGC 7789 is another unusually rich open cluster. Caroline Herschel is credited with its discovery. Estimates indicate that at least 300 stars make up the cluster, yet they remain unseen through most binoculars, as none of these stars is brighter than magnitude 10.7. Instead, we see a gentle collective blur floating amidst a field strewn with stardust.

Centaurus (Cen)

Object	Typ	R.A. (2000) h m	Dec. (2000) ° ′	Mag.	Size/Sep/ Period	Comments
NGC 3680	OC	11 25.7	−43 15	7.6	12′	
Omicron	**	11 31.8	−59 27	5.1,5.1	264″	Optical
NGC 3766	OC	11 36.1	−61 37	5.3	12′	
IC 2944	OC	11 36.6	−63 02	4.5	15′	**Lambda Cen cluster**
IC 2948	DN	11 38.3	−63 22		75′ × 50′	**Around Lambda Cen**
Stock 14	OC	11 44.0	−62 30	6.3	4′	
NGC 3909	OC	11 49.5	−48 16		30′	

(Continued on next page)

Cen (continued)

Object	Typ	R.A. (2000) h m	Dec. (2000) ° '	Mag.	Size/Sep/ Period	Comments
NGC 3918	PN	11 50.3	−57 11	8.4p	12″	
W	Vr	11 55.0	−59 15	7.6–13.7	201.6 days	Long Period Variable
Delta	**	12 08.4	−50 43	3,5,6	269″,217″	325°; 227° (1913)
XZ	Vr	12 24.2	−35 38	7.8–10.7	290.7 days	Long Period Variable
U	Vr	12 33.5	−54 40	7.0–14.0	220.28 days	Long Period Variable
NGC 4852	OC	13 00.1	−59 36	8.9p	11′	
NGC 4945	Gx	13 05.4	−49 28	9	20′ × 4′	SBc
Δ 133	**	13 22.6	−60 59	5.4,6.5	60″	343° (1879)
NGC 5128	Gx	13 25.5	−43 01	7	18′ × 14′	S0p Centaurus A
NGC 5139	GC	13 26.8	−47 29	3.7	36′	Omega Centauri
NGC 5138	OC	13 27.3	−59 01	7.6	8′	
Ru 108	OC	13 32.4	−58 29	7.5	12′	
RV	Vr	13 37.5	−56 29	7.0–10.8	446.0 days	Long Period Variable
T	Vr	13 41.8	−33 36	5.5- 9.0	90.4 days	Semi-Regular
NGC 5286	GC	13 46.4	−51 22	7.6	9′	
NGC 5281	OC	13 46.6	−62 54	5.9	5′	
NGC 5299	OC	13 50.3	−59 52		30′	
H V 124	**	13 53.5	−35 40	6.3,8.5	67″	4° (1959)
Be 146	Dk	13 57.6	−40 00		20′ × 8′	
TW	Vr	13 57.7	−31 04	8.8->12.6p	269.27 days	Long Period Variable
NGC 5460	OC	14 07.6	−48 19	5.6	25′	
R	Vr	14 16.6	−59 55	5.3–11.8	546.2 days	Long Period Variable
Lynga 2	OC	14 24.0	−61 24	6.4	12′	
NGC 5617	OC	14 29.8	−60 43	6.3	10′	
Y	Vr	14 31.0	−30 06	8.9–10.0p	180 days	Semi-Regular
Tr 22	OC	14 31.2	−61 10	7.9	7′	
V	Vr	14 32.5	−56 53	6.4–7.2	5.49 days	Cepheid
NGC 5662	OC	14 35.2	−56 33	5.5	12′	
Alpha	**	14 39.6	−60 50	0.0,11.0	131′	Alpha and Proxima Centauri

Centaurus is one of two celestial centaurs found among the stars (the other is Sagittarius, in the summer sky). From mid-northern latitudes, this centaur is seen to just skirt the southern horizon. From a more southerly vantage point, Centaurus presents many hidden glories.

NGC 3766 is found within one of the most dazzling star-filled regions in all the heavens. Through binoculars, this rich open cluster appears as a nebulous glow with only a few of its brighter stellar constituents sprinkled against it. In all, about 100 magnitude 8 to magnitude 13 stars populate the attractive group. The area surrounding NGC 3766 offers a breathtaking stellar panorama through binoculars. A slow scan across the 4° space between Omicron and Lambda Centauri displays dozens of unrelated suns that collectively put on a spectacular show.

IC 2944 is nestled among a string of four stars just south of Lambda Centauri. Two dozen stars litter this cluster's 1° diameter, with about half shining at 9th magnitude or brighter. Surrounding Lambda itself is the soft glow of **IC 2948**, a diffuse nebula. Giant binoculars fitted with twin nebula filters stand the best chance of revealing this faint cloud.

NGC 3909 is a large, scattered collection of 9th- to 13th-magnitude suns. Although measuring ½° side to side, this open cluster proves difficult to identify due to a low stellar concentration. Your best chance of capturing NGC 3909 begins by first finding the naked eye star Delta Centauri. Scan 6° to its northwest, until you come upon an asterism of six stars. The three brightest stars are labeled c^1, c^2, and c^3 Centauri, and NGC 3909 lies nearby.

Incidentally, **Delta** is itself an interesting triple star system for binocular users. The magnitude 2.8 primary star is accompanied by a magnitude 4.7 "B" star and a magnitude 6.4 "C" star. Their wide separation allows easy resolution of all three orbs through all binoculars.

NGC 3918 is one of those rare planetary nebulae that may be identified through binoculars. Observers should not expect to see much more than a point of light, as its disk is only 12″ of arc in diameter. Nonetheless, there is something about the appearance of NGC 3918 that makes it different from the neighboring stars. Perhaps it is its distinctive blue color, or it might be the slightly blurred impression we get when the other stars look crisp. Whatever the reason, try your luck with it at your next opportunity. And a special note to giant-binocular users: do you see a bright centralized point of light in the nebula? If so, congratulations, you have spotted the 11th-magnitude central star. It should be identifiable only under the clearest skies, so wait for that special evening before attempting to find it.

NGC 5128 (Figure 7.7) is a most unusual galaxy. At first glance, it would appear to be a typical elliptical galaxy, measuring 18′ × 14′. However, closer examination reveals it to be curiously bisected by a wide obscuring dark lane—anything but typical! Astronomers have determined that NGC 5128 is actually a type S0 peculiar galaxy,

Figure 7.7
Centaurus A (NGC 5128), as photographed by Martin Germano. He
used an 8-inch f/10 Schmidt-Cassegrain telescope and 103a-F
spectroscopic film for this 45-minute exposure.

but the dark absorption band remains a mystery. Adding to their
puzzlement is the unusually strong radio noise emitted from NGC
5128, dubbed Centaurus A by radio astronomers. Studies indicate
that the radio waves flow from two large, optically invisible lobes
located to either side of the visible galaxy.

Seven-power glasses show NGC 5128 as a fuzzy 7th-magnitude
"star." Giant binoculars are capable of resolving the dark lane under
good sky conditions. NGC 5128 is estimated to be about 14 million
light years away, making its true diameter about 180,000 light years.
The strange black lane has an average width of 6,000 light years.

NGC 5139 (Figure 7.8) is the finest globular cluster in the entire sky.
Better known as Omega Centauri, it is easily visible with the naked
eye as a "fuzzy" star. For ages, it was indeed thought to be just a star.
Ptolemy catalogued it as such in his catalogue of 140 A.D., as did
Bayer in his *Uranometria* atlas of 1603, where it was first assigned
the Greek letter Omega. In 1677, Halley was the first to discover the
true nature of Omega Centauri.

Astronomers estimate that Omega Centauri is 17,000 light years
from Earth, making it one of the closest globular clusters. Visually,
it subtends 36' and appears noticeably oblate.

Some of the estimated one million stars that comprise Omega
Centauri are actually resolvable in 11×80 binoculars, while a not-

Figure 7.8
The finest globular cluster of all: Omega Centauri (NGC 5139).
Photograph by Australia's Jim Barclay, F.R.A.S.

quite-resolved "grainy" appearance is evident in 7×50s. Regardless of the instrument used, an observer's first encounter with Omega Centauri will be remembered for a lifetime. Incidentally, Omega is theoretically visible from as far north as the latitude of New York City (40° north), but its image will suffer greatly owing to atmospheric interference.

T Centauri is a semiregular variable star that breaks the naked eye barrier about every three months, when it is close to magnitude 5.5. In between, its starlight falls off to 10th magnitude, and is just visible through 10× glasses. Look for the reddish gleam of T Centauri to the west of a stellar triangle formed by i, g, and k Centauri.

Also in the area is **TW Centauri**. This long period variable changes in brightness from about visual magnitude 7.5 to 13 over a 269-day period.

NGC 5281 is a dense 6th-magnitude concentration of 40 stars gathered within a small 5′ of arc diameter. The astronomer Lacaille was the first to set eyes on this 4,200-light-year-distant cluster back in 1752. A magnitude 6.6 star dominates the field of magnitude 9 and fainter luminaries. Larger glasses will also display two curving arcs of stars across the cluster's center.

NGC 5299 is a bright stellar collection found about 2° west of Hadar (Beta Centauri). It spans nearly 1° and stands out nicely against the surroundings. Recent theories indicate that NGC 5299 may not be an open cluster after all, but simply a rich Milky Way field. Regardless, it is well worth a visit.

NGC 5460 displays the largest apparent diameter of any open cluster in Centaurus. A keen-eyed observer might be able to spot this collection of two dozen stars without any optical aid at all on an extremely transparent night. With the aid of binoculars, the cluster unfolds into a misty patch of light punctuated by its brightest starry member, an 8th-magnitude orb.

R Centauri attains a greater brilliance than any other of the dozen binocular variables in Centaurus. If you faithfully monitor R over its unusually long 546-day cycle, you will find that it experiences not one, but two, maxima and minima. The minima regularly alternate between magnitudes 9 and 11 and occasionally dip as low as magnitude 13. The maxima typically register between magnitude 5.3 and magnitude 5.6.

Alpha Centauri (Rigel Kentaurus) is the brightest star within Centaurus and the third brightest star in the entire sky. Looking like a brilliant blue-white sapphire suspended in midair, Alpha Centauri is famous for being the closest star to our own.

Actually, when we look at Alpha Centauri, we are seeing a triple star system. The largest and most brilliant member of this stellar family is a spectral type-G star of magnitude −0.04. This sun is very similar in both size and nature to our own. The "B" star shines at magnitude +1.17 and orbits the "A" star once in 80 years. Their apparent separation varies from a mere 2″ to 22″ of arc over the orbital period. The third star in the system is **Proxima Centauri,** a tiny red dwarf. Collectively, the three stars lie 4.34 light years away, with Proxima slightly closer than the other two.

A clean separation of the "A" and "B" stars is possible through 10× and greater binoculars, but only at or near greatest separation. This last occurred in 1981. Each year since, and thereafter, the stars appear to draw closer together until their next closest approach, or *periastron,* in 2035. Proxima is seen as a dim magnitude 10.7 speck about 2° southwest of the brighter pair. This is too faint to be visible through most glasses.

Cepheus (Cep)

Object	Typ	R.A. (2000) h m	Dec. (2000) ° '	Mag.	Size/Sep/ Period	Comments
OΣΣ 1	**	00 14.0	+76 02	7.6,7.9	76"	103° (1923); optical
NGC 188	OC	00 44.4	+85 20	8.1	15'	
U	Vr	01 02.3	+81 53	6.7–9.2	2.493 days	Eclipsing Binary
NGC 6939	OC	20 31.4	+60 38	7.8	8'	
NGC 6946	Gx	20 34.8	+60 09	8.9	11' × 10'	Sc
B150	Dk	20 50.6	+60 18		60' × 3'	
NGC 7023	DN	21 01.8	+68 12	6.8	18' × 18'	
T	Vr	21 09.5	+68 29	5.2–11.3	388.14 days	Long Period Variable
B152	Dk	21 14.5	+61 45		15' × 3'	1° SW of Alpha Cep
S	Vr	21 35.2	+78 37	7.4–12.9	486.84 days	Long Period Variable; carbon star
IC 1396	OC/ DN	21 39.1	+57 30	3.5	50'	
B161	Dk	21 40.3	+57 49		13' × 3'	near IC 1396
Mu	Vr	21 43.5	+58 47	3.4–5.1	730 days	Semi-Regular; "Garnet star"
Hrr 11	OC	21 48	+61		600' × 300'	Includes Cep OB2 Association
NGC 7160	OC	21 53.7	+62 36	6.1	7'	
B169–71	Dk	21 58.9	+58 45		80'	3° NE of IC 1396
B173–4	Dk	22 07.4	+59 10		40'	NE of B169
NGC 7235	OC	22 12.6	+57 17	7.7	4'	
NGC 7261	OC	22 20.4	+58 05	8.4	6'	
NGC 7281	OC	22 24.7	+57 50			
Delta	Vr	22 29.2	+58 25	3.5–4.4	5.366 days	Cepheid prototype
W	Vr	22 36.5	+58 26	7.0–9.2		Semi-Regular
NGC 7510	OC	23 11.5	+60 34	7.9	4'	

According to mythology, **Cepheus** was once the mighty king of Ethiopia. After his reign ended, he was placed in the sky, along with his queen Cassiopeia and princess Andromeda, where we find them to this day. Both Cassiopeia and Andromeda are quite conspicuous in the northern hemisphere's autumn sky, but Cepheus can prove difficult to pick out. Though faint to the naked eye, Cepheus blossoms forth with over 20 deep sky objects when examined through binoculars.

NGC 188 has the dual distinction of being both the northernmost and the oldest open cluster visible from Earth. Located only 4° from Polaris, NGC 188 consists of about 120 stars ranging from 12th to 18th magnitude. Together, they merge into an 8th-magnitude glow spanning 15'. While some eagle-eyed observers have detected the dim presence of NGC 188 in 7× binoculars on extremely clear nights, the cluster's low surface brightness usually makes it a difficult find even in 11 × 80 glasses.

Studies indicate that NGC 188 is the most ancient of the known open clusters. Most of this breed are quite young on a cosmic age scale and contain many instrinsically brilliant blue giant stars. As a cluster ages, these short-lived suns consume their fuel rapidly. Older clusters, such as NGC 188, are nearly devoid of blue giants. The best estimates place the age of NGC 188 at 5 billion years, as compared with 190 million years for M41 in Canis Major and only 4 million years for the Double Cluster in nearby Perseus.

NGC 6939 and **NGC 6946** are both tucked just inside the Cepheus border, about 3° southwest of Eta Cephei. The former is a fairly bright but small open cluster that defies resolution through binoculars. Though their light combines to an overall 8th-magnitude, none of this cluster's 80 stars shine brighter than 12th magnitude. With so many stars crammed into a tiny 8'-of-arc diameter, NGC 6939 is one of the richest clusters in the northern autumn sky.

While NGC 6939 lies about 4,000 light years away, its "neighbor" is over 16 million light years from Earth. A magnificent face-on Sc spiral galaxy, NGC 6946 is one of Cepheus' most challenging objects to find. It is likely to remain invisible through most glasses, and even giant binoculars will show an exceedingly faint, nearly circular grayish patch of light highlighted by an ever-so-slightly brighter central core.

IC 1396 is a huge region of emission nebulosity measuring nearly 1° in diameter. Due to its wide expanse, few amateur telescopes have a broad enough field to take it all in. Binoculars, however, thanks to their wider view, are capable of revealing this delicate cloud. Sightings of IC 1396 have been reported in 7 × 50 glasses equipped with contrast-enhancing nebula filters, while in 15× binoculars the cloud appears as a broken, irregular wreath of grayish light embedded with several centrally located stars. Without the aid of nebula filters, IC 1396 will remain unseen.

Mu Cephei is located on the northeastern edge of IC 1396. Known as Herschel's Garnet Star, Mu is one of the reddest stars in the entire sky that is visible through binoculars. It shines like a striking deep orange or red beacon amidst an infinite sea of stars.

A further enticement for observers to visit Mu often is in its semi-regular light fluctuations. Over a period of approximately 730 days, Mu varies between magnitudes 3.4 and 5.1. Studies indicate secondary oscillations with periods ranging from 100 to 4,500 days. Mu is classified as a type-M pulsating red giant, similar to Betelgeuse in Orion, with an intrinsic luminosity perhaps 12,000 times greater than our sun!

Harrington 11 appears as a conspicuous bright band of starlight that breaks off from the mainstream of the Cygnus Milky Way and veers northward into the southwestern portion of Cepheus. To the unaided eye, it measures about 10° × 5° and looks like a perfectly straight, detached section of our galaxy. Formed in part by the **Cepheus OB2** stellar association, Harrington 11 resolves into a myriad of separate stars through binoculars. Held within its somewhat jagged borders are several deep sky objects, including Mu Cephei, IC 1396, NGC 7160, and NGC 7235. Together with Harrington 10, a parallel dark band that crosses the Milky Way about 8° to the south in Cygnus, they form a unique pair of easy, though little noticed, objects.

Delta Cephei is the prototype sun of a family of yellow giant variable stars that has contributed to our present understanding of the size and distances in the universe more than any other single type of star. Cepheid variables exhibit well-documented relationships between their periods of magnitude modulation and intrinsic brightness (absolute magnitude). By comparing a star's absolute magnitude with its observed apparent magnitude, astronomers can calculate the star's distance. Cepheids, being giant stars, have also been observed in several nearby galaxies, thereby permitting precise distance calculations to these galactic systems as well.

The variability of Delta Cephei was first noted by British astronomer John Goodricke in 1784. Since then, Delta has been monitored almost continuously. Its brightness varies from magnitude 3.5 to 4.4 and has a measured period of 5 days, 8 hours, and 47 minutes.

Cetus (Cet)

Object	Typ	R.A. (2000) h m	Dec. (2000) ° '	Mag.	Size/Sep/ Period	Comments
W	Vr	00 02.1	−14 41	7.1–14.8	351.31 days	Long Period Variable
T	Vr	00 21.8	−20 03	5.0–6.9	158.9 days	Semi-Regular
NGC 246	PN	00 47.0	−11 53	8.5p	240″ × 210″	
NGC 247	Gx	00 47.1	−20 46	8.9	18′ × 5′	S-
37	**	01 14.4	−07 55	5.2,8.7	50″	331° (1931); 1003
UV	Vr	01 38.8	−17 58	6.8–13.0		Irregular

(Continued on next page)

Cet (continued)

Object	Typ	R.A. (2000) h m	Dec. (2000) ° '	Mag.	Size/Sep/ Period	Comments
Omicron	Vr	02 19.3	−02 59	2.0–10.1	331.96 days	Long Period Variable; "Mira"
R	Vr	02 26.0	−00 11	7.2–14	166.24 days	Long Period Variable
U	Vr	02 33.7	−13 09	6.8–13.4	234.76 days	Long Period Variable
NGC 1068	Gx	02 42.7	−00 01	8.9	6' × 5'	Sbp M77; Seyfert galaxy

Cetus, the Whale, spans a wide region of our autumn sky from Aquarius and Pisces in the west to Taurus in the east. To those familiar with the ancient Greek legend of Andromeda and Perseus, Cetus is the sea monster sent by Poseidon to devour Andromeda. In the end, she is saved just in time by Perseus.

The astronomical Cetus is best known as the home of the famous long period variable Mira, but it also holds several other variables, a double star, a planetary nebula, and two galaxies all within range of binoculars.

NGC 246 is one of the largest planetaries in the sky, but its low surface brightness makes it one of the most difficult to spot. Weighing in at photographic magnitude 8.5, its ghostly 4'-diameter shell is just detectable in large nebula-filtered binoculars as a slightly oval glimmer. Photographs show that NGC 246 has an irregular ringlike structure around a 12th-magnitude central star and three unrelated dim suns.

NGC 247 rides close to the southern border of Cetus, about 3° south of Beta Ceti. Belonging to the nearby Sculptor Galaxy Group, NGC 247 is marginally visible in 10 × 70 and larger glasses as a cigar-shaped smudge with a condensed central nucleus. Although classified as a spiral, this galaxy reveals a unique, mottled appearance in photographs that is seen in few other galaxies.

Omicron Ceti, better known as "Mira" ("the Wonderful"), is the prototype of the "long period variable" family of stars. Long period variables are symmetry in motion, as they slowly and predictably fluctuate between maximum and minimum brightnesses. All are thought to be ancient red giant stars that actually pulsate over time due to internal changes.

First detected as a variable in 1596 by German astronomer David Fabricius, Mira is an ideal star for first-time variable star observers.

It typically cycles from 3rd to 10th magnitude over a 332-day period, but has been known to approach 2nd magnitude on occasion. Chapter 6 gives a finder chart and further description of Mira.

M77 (NGC 1068) is the brightest of several galaxies found near Delta Ceti and the only one likely to be seen through binoculars. Look for its small, rather faint disk about halfway between Delta and an isosceles triangle of 8th-magnitude stars 1½° to the southeast. Binoculars reveal the bright central core of the galaxy, but its face-on spiral arms can be seen in relatively large telescopes only.

Although indexed as an Sb spiral galaxy, M77 is also known to be a *Seyfert galaxy*. Seyfert galaxies, named for their discoverer, American astronomer Carl Seyfert, exhibit unusually intense and variable emissions at ultraviolet wavelengths.

M77 also carries its own footnote in astronomical history books as being one of the first galaxies in which a large redshift was detected. This 1913 discovery by Vesto Slipher led to the expanding universe theory that remains popular to this day.

Chamaeleon (Cha)

Object	Typ	R.A. (2000) h m	Dec. (2000) ° '	Mag.	Size/Sep/ Period	Comments
R	Vr	08 21.8	−76 21	7.5–14.2	334.58 days	Long Period Variable
Delta[1+2]	**	10 45.3	−80 28	4.5,5.5	360"	Optical

Chamaeleon, the Chameleon, is a small, unimpressive constellation that lies too far south to be seen from the United States or any European country. Only when situated below 10° north latitude will observers readily be able to find its few naked eye stars. Even then, its location far from the plane of our galaxy provides few sights of interest.

R Chamaeleonis is a long-period variable found a degree north of Alpha Chamaeleonis. It varies from magnitude 7.5 to magnitude 14.2 over a 334-day cycle and is set very near a 7th-magnitude star of fixed brightness. Be sure not to confuse the fixed star with the variable, especially when the latter is below the visibility threshold of binoculars.

Delta[1] and **Delta[2]** Chamaeleonis form an easily split pair of 5th-magnitude stars. Though not actually close to each other in space, their chance alignment as seen from Earth provides a pleasant target for even the smallest opera glasses. Delta[1] appears pure white, while fainter Delta[2] exhibits a yellowish glint.

Circinus (Cir)

Object	Typ	R.A. (2000) h m	Dec. (2000) ° '	Mag.	Size/Sep/ Period	Comments
Δ 169	**	14 45.2	−55 36	6.2,7.6	68″	106° (1938)
NGC 5823	OC	15 05.7	−55 36	7.9	10′	
Pismis 20	OC	15 15.4	−59 04	7.8	5′	
Gamma	**	15 20.4	−59 19	4.5,8	180″	Optical

Circinus, the Compasses, is a small constellation found to the south and east of the bright stars Alpha and Beta Centauri. First named in 1752 by Lacaille, it is immersed in the hazy band of our galaxy's plane, yet holds few targets of interest to observers with binoculars.

NGC 5823 lies just within the northern border of Circinus. It is a rich open cluster embodying some 100 stars shining at 10th magnitude and fainter and spanning 10′ of arc. Current estimates place NGC 5823 at about 2,300 light years away.

·**Gamma Circini** forms an attractive and colorful pair with an 8th-magnitude orb found about 3′ to its northeast. Gamma shines with a yellowish glint, while the faint "B" star appears reddish through large glasses. Although these two stars are not actually related to each other, Gamma itself is a true binary system, with a 5th-magnitude companion located only 1″ of arc away.

Columba (Col)

Object	Typ	R.A. (2000) h m	Dec. (2000) ° '	Mag.	Size/Sep/ Period	Comments
NGC 1851	GC	05 14.1	−40 03	7.3	11′	X-ray source
T	Vr	05 19.3	−33 42	6.6–12.7	225.84 days	Long Period Variable
h 3849	**	06 19.8	−39 29	6.7,8.3	40″	53° (1950)
h 3857	**	06 24.0	−36 42	5.7,6.9	64″	72° (1960)

Columba, the Dove, is a faint southern constellation wedged between the far more brilliant regions of Canis Major and Puppis. Its origin dates back to 1603, when it was first depicted on Bayer's star charts as a dove near the mighty ship Argo Navis. While the monstrous constellation Argo has since been sliced into separate constellations, little Columba remains untouched. Only four deep sky objects are found with binoculars inside its diminutive borders.

NGC 1851 is the only nonstellar deep sky object found in Columba through binoculars. Due to its isolation from any bright stars, observers should be prepared to search for a while. Once spotted, however, this magnitude 7.3 globular cluster stands out quite well as a small, circular patch of fuzzy light. Take a careful look at its core; astronomers have detected intense X rays coming from within, indicating that there must be more there than meets the eye.

Coma Berenices (Com)

Object	Typ	R.A. (2000) h m	Dec. (2000) ° '	Mag.	Size/Sep/ Period	Comments
R	Vr	12 04.0	+18 49	7.1–14.6	362.8 days	Long Period Variable
NGC 4254	Gx	12 18.8	+14 25	9.8	5'	Sc M99
NGC 4321	Gx	12 22.9	+15 49	9.4	7' × 6'	Sc M100
Mel 111	OC	12 25	+26	1.8	275'	Coma star cluster
NGC 4382	Gx	12 25.4	+18 11	9.2	7' × 5'	Ep M85
17	**	12 28.9	+25 55	5.3,6.6	145"	251° (1928); 8568 (in Mel 111)
NGC 4501	Gx	12 32.0	+14 25	9.5	7' × 4'	Sb+ M88
NGC 4559	Gx	12 36.0	+27 58	9.9	10' × 5'	Sc
NGC 4565	Gx	12 36.3	+25 59	9.6	16' × 3'	Sb
NGC 4725	Gx	12 50.4	+25 30	9.2	11' × 8'	SBb
32+33	**	12 52.2	+17 04	6.3,6.7	95"	49° (1922)
NGC 4826	Gx	12 56.7	+21 41	8.5	9' × 5'	Sb- M64; Black-Eye galaxy
NGC 5024	GC	13 12.9	+18 10	7.7	13'	M53

Ptolemy III ruled ancient Egypt in the middle of the third century B.C. According to legend, his queen, Berenice, was well known for her long, beautiful hair. One day, as her husband rode off into battle, Berenice vowed to the goddess Aphrodite that she would sacrifice her flowing locks if her husband was permitted to come back to her safely. Upon his return, she kept her promise by cutting off her hair and presenting it to the gods. But the hair mysteriously disappeared. Conon, the court astronomer, pointed to the night sky, toward what is now our constellation **Coma Berenices,** and assured the king and queen that Aphrodite had placed the queen's hair among the stars for all the world to see.

M99 (NGC 4254) is a difficult galaxy to catch through binoculars. Glowing dimly at 10th magnitude, its round disk is located about $7\frac{1}{2}°$ due east of Denebola (Beta Leonis). Look for a row of three 5th-magnitude stars, and then drop south about $\frac{1}{2}°$, where M99 resides next to a 7th-magnitude star. If you successfully capture M99, then try your luck with an even more elusive galaxy. Located nearby, **M100 (NGC 4321)** is a half-magnitude brighter than M99, which might lead an observer into expecting it to be an easier target. However, due to its larger apparent diameter, the surface brightness of this Sc spiral is quite low, making it the more difficult of the pair to discern. Both of these galaxies will undoubtedly require giant glasses to confirm their presence.

M99 and M100 are two of the many galaxies in this region that belong to the Virgo Realm of Galaxies. The Realm of Galaxies is the closest galaxy cluster to the Milky Way and claims hundreds of galactic associates. Large binoculars can reveal a few of the brighter ones, but the majority remain strictly telescopic objects.

Melotte 111 (Figure 7.9), popularly known as the Coma Berenices Star Cluster, is what Conon saw as his queen's heavenly tresses. In reality, this is an impressive collection of 80 5th- and 6th-magnitude stars spanning almost 5° of sky. Collectively, the group is some 260 light years away and is estimated to be about 400 million years old. Due to the large expanse of sky covered by the cluster, low-power, wide-field binoculars are the best instruments for viewing Melotte 111. Highly magnified, narrow-field instruments only cause the clustering effect to be lost.

While there are several double stars within Melotte 111, the only one easily resolvable in binoculars is **17 Comae Berenices.** Found a bit to the east of the cluster's center, 17 consists of two type-A suns. Watch for the 6th-magnitude "B" star about $2\frac{1}{2}'$ west-southwest of the 5th-magnitude primary.

M85 (NGC 4382) is one of the Realm's brighter galaxies. Méchain was the first to come upon its magnitude 9.2 glow in 1781. Though only faintly visible through 7 × 50 glasses on dark nights, it is a fairly easy target to snare in 11 × 80 binoculars, even from light-polluted

Figure 7.9
The Coma Berenices Star Cluster (Melotte 111), as photographed by Lee
Coombs. He used a 50-mm lens stopped down to f/2 for this two-minute
exposure on ISO 64 film.

suburban skies. A brighter stellar nucleus is seen surrounded by a
fainter halo.

M88 (NGC 4501) is another galaxy that is not too difficult to spot.
Unfortunately, guide stars in its immediate vicinity are few and far
between. Once you zero in on M88, you will be looking at an Sb+
spiral galaxy whose profile is tilted 30° to our line of sight. Binoculars
reveal a moderately bright core engulfed by the spiral arms' faint
glow. Overall, M88 spans about 7′×4′ of arc.

NGC 4565 is one of the most impressive examples of an edge-on
spiral galaxy found anywhere in the visible universe. Through 10×

and larger glasses, it appears as a fine spindle of light floating against a starry backdrop. The bulge of its central hub, so striking in photographs, is just discernible in 20× binoculars.

M64 (NGC 4826), nicknamed the Black-Eye Galaxy, is Coma Berenices' brightest galaxy by far. It was discovered in 1775 by J. E. Bode, who described it as a "small, nebulous star." Shining at magnitude 8.5, it may be glimpsed with 7× as an oval patch of light with a brighter center. Experienced observers have reported spotting its mysterious "black eye" dust lane through 11 × 80's on clear, moonless nights.

M53 (NGC 5024), a globular cluster also discovered by Bode in 1775, seems a bit out of place this far from the Milky Way's plane. Binoculars display a round, nebulous disk drawing to a brighter center. Look for M53 as a 13'-diameter puff set in an attractive star field. None of the cluster's 100,000 or so stars is bright enough to be seen through binoculars.

Corona Australis (CrA)

Object	Typ	R.A. (2000) h m	Dec. (2000) ° '	Mag.	Size/Sep/ Period	Comments
NGC 6541	GC	18 08.0	−43 42	6.6	13'	
AM	Vr	18 41.2	−37 29	8.6–12.7p	187.5 days	Semi-Regular
NGC 6726	DN	19 01.7	−36 53		2' × 2'	
Be 157	Dk	19 02.9	−37 08		55' × 18'	SE of NGC 6726
SL 42	Dk	19 10.3	−37 08		12' × 8'	2° E of NGC 6726

Corona Australis, the Southern Crown, is seen as an arc of stars immediately to the south of the Sagittarius teapot. Positioned along the southern Milky Way, Corona Australis offers many attractive star fields for the cruising binocularist to pause, as well as a bright globular cluster and a selection of challenging bright and dark nebulae.

NGC 6541 is the brighter of two globular clusters in the Southern Crown. (The other, NGC 6496, is not readily visible in binoculars.) Located amid an attractive stellar setting just south of a 5th-magnitude star, NGC 6541 appears as a not-quite-stellar point of light through 7× glasses. Giant binoculars reveal a distinct disk, but resolution of its stars is impossible.

NGC 6726 is the brightest tip of a huge cloud of nebulosity that covers much of northeastern Corona Australis. It may be found surrounding an 8th-magnitude star just west of Gamma Coronae Australis. Nebula hunters will find it a difficult test even through giant glasses, as the light from Gamma tends to flood the field and obliterate the delicate cloud.

Bernes 157 is a large, irregularly shaped dark cloud that bridges the 1° gap between NGC 6726 and Gamma Coronae Australis. Its silhouette is apparent through binoculars due to a total absence of any detectable foreground stars. **Sandqvist and Lindroos 42,** a second dark cloud, lies a degree to the east of Gamma, but is less conspicuous due to its much smaller apparent size.

Corona Borealis (CrB)

Object	Typ	R.A. (2000) h m	Dec. (2000) ° '	Mag.	Size/Sep/ Period	Comments
U	Vr	15 18.2	+31 39	7.7–8.8	3.452 days	**Eclipsing Binary**
S	Vr	15 21.4	+31 32	5.8–14.1	360.26 days	**Long Period Variable**
SW	Vr	15 40.8	+38 43	7.8–8.5	100 days	Semi-Regular
RR	Vr	15 41.4	+38 33	8.4–10.1	60.8 days	Semi-Regular
R	Vr	15 48.6	+28 09	5.7–14.8		**Irregular; R CrB prototype**
V	Vr	15 49.5	+39 34	6.9–12.6	357.63 days	Long Period Variable
RS	Vr	15 58.5	+36 01	8.7–11.6p	332.2 days	**Semi-Regular**
T	Vr	15 59.5	+25 55	2.0–10.8	29,000 days?	**Recurrent nova** (1946)
H V 38	**	16 22.9	+32 20	6.3,8.8	35″	19° (1914); 10031

To the east of Boötes is an attractive semicircle of stars known as **Corona Borealis,** the Northern Crown. Ptolemy is credited with devising this constellation, although it had been previously known for some time as a wreath. The only objects within the Northern Crown that are of interest to viewers with binoculars are a few variable and double stars. Many galaxies dot Corona Borealis, but all are restricted to large telescopes.

S Coronae Borealis is found to the west of Theta Coronae Borealis, the westernmost star in the crown's arc. A 7th-magnitude sun lies just to the variable's northeast and helps observers pinpoint the exact location of this 360-day long-period star. At maximum intensity, S Coronae Borealis outshines this fixed-brightness star by about a full magnitude, but it fades to 14th magnitude at minimum intensity. The eclipsing binary **U Coronae Borealis** lies about a degree farther northwest of S Coronae Borealis. Over its 3.452-day cycle, U Coronae Borealis fluctuates from magnitude 7.7 to magnitude 8.8.

R Coronae Borealis is a most intriguing variable star indeed. "R Cor Bor," as it is affectionately called by ardent variable star observers, typically remains about 6th magnitude, but will suddenly plummet toward 14th magnitude in just a few weeks. Most of these unannounced descents in brightness last for several months, although some have continued for years on end. Then, as unpredictably as it faded, R will ascend from the depths of obscurity back to its usual brilliance. This odd behavior has been likened to a nova in reverse and apparently is caused by a large obscuring carbon cloud emitted from the star, blocking its light. As the cloud dissipates, the star's apparent luminosity returns.

T Coronae Borealis is another odd variable found within the Northern Crown. For the most part, this star leads an unspectacular life at 10th magnitude. All this has changed abruptly on at least two occasions in the past, as T has erupted to about 2nd or 3rd magnitude. On May 12, 1866, it peaked at magnitude 2.0. The star then fell to magnitude 9.5 in less than a month, and settled back to 10th magnitude in about 250 days. Eighty years later, it flared again to 3rd magnitude. Soon, it again faded to 10th magnitude, and it has remained there ever since. When it will burst again, no one really knows.

H V 38, included in most astronomical lists as 23 Herculis, was originally assigned to Hercules, but now finds itself with Corona Borealis. Large glasses are suggested for the best view, although the components of this binary system may be separated using small binoculars under superior seeing conditions. Look for a magnitude 8.8 secondary star adjacent to the 6th-magnitude primary sun. They are separated by only 35" of arc, with each shining pearly white.

Corvus (Crv)

Object	Typ	R.A. (2000) h m	Dec. (2000) ° '	Mag.	Size/Sep/ Period	Comments
R	Vr	12 19.6	−19 15	6.7–14.4	317.03 days	Long Period Variable
NGC 4361	PN	12 24.5	−18 48	10.3p	45″	Giant glasses only

Legend tells us that **Corvus,** the Crow, once belonged to mighty Apollo. Today, we find Corvus represented by a trapezoid of 3rd-magnitude stars set in an empty region of the southern spring sky. Using binoculars, most sky enthusiasts will find only a lone target within.

R Corvi is found nearly centralized in the trapezoid. It is easily spotted as one of three orbs that form a tight stellar triangle. Over the course of its 317-day cycle, R will outshine the other two stars when it is near its maximum magnitude of 6.7, only to disappear from view as it heads toward a minimum magnitude of 14.4.

Crater (Crt)

Object	Typ	R.A. (2000) h m	Dec. (2000) ° '	Mag.	Size/Sep/ Period	Comments
RU	Vr	11 51.1	−11 12	8.5–9.5p		Irregular

Crater, the Cup, is found alongside Corvus in the spring sky, riding on the back of Hydra. Its form and identity originated in Ptolemy's *Almagest* and has been associated with several mythological gods and heroes. Observers with binoculars will find little of interest within Crater; a single variable star is all that attracts our attention. Many galaxies also dot the area, but are far below binocular visibility.

RU Crateris is found just above the "rim" of the Cup marked by Theta and Eta Crateris. Observers will have to look long and hard to detect this irregular variable, as it never shines above photographic magnitude 8.5. However, since it also never dips below magnitude 9.5, it may be continually monitored through 10× and larger instruments.

Crux (Cru)

Object	Typ	R.A. (2000) h m	Dec. (2000) ° '	Mag.	Size/Sep/ Period	Comments
Ru 98	OC	11 58.0	−64 29	7.0	10'	
NGC 4052	OC	12 01.9	−63 12	8.8p	8'	
NGC 4103	OC	12 06.7	−61 15	7.4p	7'	
BH	Vr	12 16.3	−56 17	7.2–10.0	421 days	Long Period Variable
AO	Vr	12 17.8	−63 37	8.5–10.0p		Irregular
R	**Vr**	12 23.6	−61 38	6.4–7.2	5.83 days	**Cepheid**
NGC 4337	OC	12 23.9	−58 08	8.9	4'	
NGC 4349	OC	12 24.5	−61 54	7.4	16'	
Alpha	**	12 26.6	−63 06	1.4,4.9	90"	202° (1913); Acrux
NGC 4439	OC	12 28.4	−60 06	8.4	4'	
Ru 165	OC	12 28.7	−56 28		22'	
Harvard 5	OC	12 29.0	−60 46	7.1	6'	
Gamma	**	12 31.2	−57 07	1.6,6.7	111"	31° (1919)
Harvard 7	OC	12 39.7	−60 36		8'	
NGC 4609	OC	12 42.3	−62 58	6.9	5'	
Coalsack	Dk	12 53	−63		400' × 300'	
NGC 4755	OC	12 53.6	−60 20	4.2	10'	Jewel Box
Mu	**	12 54.6	−57 11	4.3,5.3	35"	17° (1952)

What a marvelous constellation **Crux** is! Although it is the smallest of the 88 constellations, the Southern Cross (as it is popularly known) is one of the richest stellar fields anywhere in the heavens. Visible well only from southern latitudes, some of the heaven's finest open clusters as well as the largest dark nebula visible from Earth are located within its borders.

Ruprecht 98 is found in the extreme southwestern corner of Crux, about a degree west of 4th-magnitude Eta Crucis. Although Ruprecht 98 was passed over in the *New General Catalogue*, binoculars can discern this tiny open cluster as a 7th-magnitude glow buried among the stars of the southern Milky Way. Larger glasses will disclose some of the more illustrious of its 50 stars.

NGC 4052 is situated in an enchanting star-filled region just north-west of the bright optical double, Theta^1 and Theta^2 Crucis. Though not as prominent as some of the other clusters in Crux, it may be spotted as a fairly bright object engulfed in a starry setting. Estimates indicate that 80 suns belong to NGC 4052 and that they lie about 6,200 light years from the Solar System.

NGC 4103 is a highly concentrated swarm of 45 stars located just south of a pair of 6th-magnitude suns. Ten-power and greater binoculars can pick out a few 9th-magnitude points of light amidst the cluster's hazy mass. The region is perfect for a leisurely stellar stroll with any pair of quality binoculars.

R Crucis is an easily observable Cepheid-type variable star. Fluctuating between magnitudes 6.4 and 7.2 over a period slightly less than six days, it is a perfect variable for study through binoculars.

NGC 4349 is a dense galactic cluster situated between R Crucis and a small asterism of 6th- and 7th-magnitude suns. With a visual magnitude of 7.4, NGC 4349 is readily detectable in most glasses as a misty patch of stardust along the plane of our galaxy. Resolution of the cluster stars is all but impossible through binoculars, since the brightest are only 11th magnitude.

Alpha Crucis is the southernmost of the four stars forming the Southern Cross. **Acrux,** as it is popularly known, looks like a brilliant blue sapphire encircled by a myriad of faint stellar jewels.

Telescopes reveal Acrux to be a striking binary star, with a magnitude 1.4 primary sun paired with a 2nd-magnitude companion. Unfortunately, the pair is separated by only 4.4" of arc, which is too close to resolve through binoculars. However, Alpha Crucis C, a third member of the system, is clearly visible as a 5th-magnitude star about 90" from the brighter pair. Like the two brighter components of Acrux, Alpha C is a searing type-B star.

Gamma Crucis marks the top of the Southern Cross and is another striking beacon of the southern sky. Gamma, with a distinctly orange tint, forms an apparent pair with whitish, magnitude 6.7 Gamma B to its southeast. Nearly 2' of arc separate this colorful duo.

NGC 4609 is a small but pleasing galactic cluster. Consisting of about 40 magnitude 9 and fainter stars, NGC 4609 stands out well against the almost starless backdrop of the famous Coalsack. Only an adjacent 5th-magnitude point of light is visible.

The **Coalsack** (Figure 7.10) is the most famous naked-eye dark nebula in the entire sky. Covering an area approximately 5° × 7°, it looks like a huge dark hole in the radiant star clouds of the Milky Way just east of Acrux. It more than fills the field of most binoculars. While most amateurs see the Coalsack as a homogeneous black patch, intricate filamentary detail may be spotted throughout by sharp-eyed

Figure 7.10
The Coalsack dominates this view of Crux and the southern Milky Way.
Photograph by Jack Newton.

observers. However, one must wait for a dark night before these delicate features will reveal themselves. At a distance estimated to be about 550 light years away, the 60-light-year-wide Coalsack is believed to be the closest obscuring nebulosity to Earth.

NGC 4755, a marvelous open cluster, is located on the northern edge of the Coalsack. Referred to by many as the Kappa Crucis Cluster after its most eminent constituent, NGC 4755 is best known as the Jewel Box Cluster for its gemlike quality. The Jewel Box ranks as one of the finest clusters in the sky, although many amateurs are surprised at first by its small size, only 10' of arc across. Nonetheless, the unparalleled beauty of NGC 4755 more than makes up for its tiny girth.

NGC 4755's stars are set in a pattern resembling an arrowhead, with the three brightest found in a row stretching from northeast to southwest. Kappa Crucis, the southwestern star, is an orange-red beacon. Another half-dozen stars ranging from magnitude 6 to magnitude 10 may be identified through 7× to 10× glasses. Many exhibit contrasting tints of color, which provide further testimony to the

Jewel Box's magnificence. Recent estimates place NGC 4755 at a distance of 7,600 light years away, although this figure may be inaccurate due to heavy obscuration from the Coalsack.

Cygnus (Cyg)

Object	Typ	R.A. (2000) h m	Dec. (2000) ° '	Mag.	Size/Sep/ Period	Comments
CH	Vr	19 24.5	+50 14	6.4–8.7	97 days	Z And type
OΣΣ 182	**	19 26.8	+50 09	7.3,8.5	73"	300° (1956); 12470
AF	Vr	19 30.2	+46 09	7.4–9.4p	94.1 days	Semi-Regular
Beta	**	**19 30.7**	**+27 58**	**3.1,5.1**	**34"**	**54° (1967); 12540; Albireo**
V1125	Vr	19 31.8	+31 52	9.0–9.9p		Irregular
PK 64+5.1	PN	19 34.8	+30 31	9.6p	8"	Campbell's Star
R	**Vr**	**19 36.8**	**+50 12**	**6.1–14.2**	**426.44 days**	**Long Period Variable**
NGC 6811	OC	19 38.2	+46 34	6.8	13'	
NGC 6819	OC	19 41.3	+40 11	7.3	5'	
RT	Vr	19 43.6	+48 47	6.4–12.7	190.28 days	Long Period Variable
NGC 6826	**PN**	**19 44.8**	**+50 31**	**9.8p**	**30" × 140"**	**Blinking Planetary**
Chi	**Vr**	**19 50.6**	**+32 55**	**3.3–14.2**	**406.93 days**	**Long Period Variable**
NGC 6834	OC	19 52.2	+29 25	7.8	5'	
B144	Dk	19 59	+35		360' × 180'	Fish-on-the-Platter Nebula
Z	Vr	20 01.4	+50 03	7.4–14.7	263.69 days	Long Period Variable
B145	Dk	20 02.8	+37 40		35' × 6'	
NGC 6866	OC	20 03.7	+44 00	7.6	7'	
NGC 6871	**OC**	**20 05.9**	**+35 47**	**5.2**	**20'**	
Basel 6	OC	20 06.8	+38 21	7.7	14'	
Biur 2	OC	20 09.2	+35 29	6.3	13'	
Roslund 5	**OC**	**20 10.0**	**+33 46**		**45'**	
NGC 6888	DN	20 12.0	+38 21		20' × 10'	Crescent Nebula
RS	Vr	20 13.4	+38 44	6.5–9.3	417.39 days	Semi-Regular
B343	Dk	20 13.5	+40 16		10' × 5'	
Omicron¹	**	20 13.6	+46 44	4,7,5	107",338"	173°, 338° (1926); 13554
IC 4996	OC	20 16.5	+37 38	7.3	6'	
P	Vr	20 17.8	+38 02	3.0–6.0	—	S Dor type
CN	Vr	20 17.9	+59 48	7.3–14.0	198.48 days	Long Period Variable
Cr 419	OC	20 18.1	+40 43	5.4p	5'	
U	Vr	20 19.6	+47 54	5.9–12.1	462.40 days	Long Period Variable
Berk 86	OC	20 20.4	+38 42	7.9	8'	
OΣΣ 207	**	20 22.9	+42 59	6.6,8.5	93"	63° (1920); 13786
NGC 6910	OC	20 23.1	+40 47	7.4	8'	
NGC 6913	**OC**	**20 23.9**	**+38 22**	**6.6**	**7'**	**M29**

(Continued on next page)

Cyg (continued)

Object	Typ	R.A. (2000) h m	Dec. (2000) ° '	Mag.	Size/Sep/ Period	Comments
LDN 889	Dk	20 24.8	+40 10		90' × 20'	
LDN 906	Dk	20 40	+42			Northern Coalsack
Ru 173	OC	20 41.8	+35 33		50'	
X	Vr	20 43.4	+35 35	5.9–6.9	16.39 days	Cepheid
NGC 6960	DN	20 45.7	+30 43		70' × 6'	Filamentary Nebula (52 Cyg)
B350	Dk	20 49.1	+45 53		3'	
IC 5067	DN	20 50.8	+44 21		80' × 70'	Pelican Nebula
NGC 6992	DN	20 56.4	+31 43		60' × 8'	Veil Nebula
LDN 935	Dk	20 56.8	+43 52		150' × 40'	Between North American and Pelican Nebulae
B352	Dk	20 57.1	+45 54		20' × 10'	
NGC 7000	DN	20 58.8	+44 20		120' × 100'	North American Nebula
Hrr 10	Dk	21 00	+55		600' × 180'	Dark lane across Milky Way
Cr 428	OC	21 03.2	+44 25	8.7p	14'	
61	**	21 06.9	+38 45	5.2,6.0	321"	195° (1976); 14636
NGC 7039	OC	21 11.2	+45 39	7.6	25'	
IC 1369	OC	21 12.1	+47 44	6.8	4'	
B361	Dk	21 12.9	+47 22		17'	
V1070	Vr	21 22.8	+40 56	6.7–7.7		Semi-Regular
NGC 7062	OC	21 23.2	+46 23	8.3	7'	
B362	Dk	21 24.0	+50 10		15' × 8'	
NGC 7063	OC	21 24.4	+36 30	7.0	8'	
NGC 7082	OC	21 29.4	+47 05	7.2	25'	
NGC 7086	OC	21 30.5	+51 35	8.4	9'	
NGC 7092	OC	21 32.2	+48 26	4.6	32'	M39
W	Vr	21 36.0	+45 22	6.8–8.9	126.26 days	Semi-Regular
RU	Vr	21 40.6	+54 19	9.2–11.6p	233.85 days	Semi-Regular
V460	Vr	21 42.0	+35 31	5.6–7.0		Irregular
V1339	Vr	21 42.1	+45 46	5.9–7.1	35 days	Semi-Regular
B164	Dk	21 46.5	+51 04		12' × 6'	
B168	Dk	21 53.2	+47 12		100' × 10'	Cocoon Nebula at east end
IC 5146	DN/ OC	21 53.5	+47 16		12' × 12'	Cocoon Nebula

Flying high overhead in northern hemisphere skies during the months of June through September is the spectacular constellation **Cygnus**, the Swan. With the summer Milky Way passing through, Cygnus contains a great variety of deep sky splendors to entice the observer. Regardless of the type and size of your binoculars, Cygnus holds something of interest.

Albireo (Beta Cygni), marking the swan's beak, is one of the summer sky's most radiant stellar jewels. Through low-power binoculars or finderscopes, sharp-eyed observers are able to discern that it is not a single star at all, but rather a pair of distant suns set against a glorious Milky Way field. These two stars are famous for their outstanding color contrast. Albireo A, a type-K star, is a gleaming yellow 3rd-magnitude gem that shines with the golden radiance of a fiery ember. Its 5th-magnitude companion, Albireo B, is a type-B star and glows a dazzling azure, much as a sapphire does when it catches a single ray of light. The colors appear even more vivid if you first softly defocus the stellar images.

R Cygni, when at or near maximum brightness, is one of the most strikingly colorful variables through binoculars. Located just a few minutes of arc to the east of Theta Cygni, R fluctuates between magnitudes 6.1 and 14.2 across a 426-day period. The deep ruddy color of this spectral type-S variable contrasts beautifully with Theta, a type-F white star that takes on a bluish tint when compared to R.

NGC 6826, found just west of 16 Cygni, is a relatively easy planetary nebula to identify through binoculars. Known popularly as the "Blinking Planetary," it shines at about 9th magnitude with a bluish glint. Giant glasses may even reach the 11th-magnitude central star. The unusual nickname comes from the nebula's visual appearance: when viewed directly, the nebula seemingly disappears, only to reappear when it is glimpsed out of the corner of your eye. Averted vision reveals NGC 6826 as a tiny circular glow.

Chi Cygni is a long period variable star noted for its broad range of magnitudes. Binoculars are the instrument of choice to follow this star as it approaches maximum. At greatest light, Chi may reach magnitude 3.3 and become visible to the naked eye. Its light then fades at a fairly constant rate until, some 200 days later, it bottoms out at about 14th magnitude. Then it begins to cycle back toward another maximum. Look for Chi along the swan's neck, approximately one-fourth of the way from Eta Cygni to Albireo.

NGC 6871 is an attractive open cluster that is easily visible in even the smallest instruments as a brighter patch of Milky Way between Gamma and Eta Cygni. Most 7× and larger binoculars reveal about eight stars between 7th and 9th magnitude set in an arc and surrounded by the faint glow of other cluster members. In all, about 15 suns down to 12th magnitude populate this $\frac{1}{2}°$ swarm.

Roslund 5 is a large, rather obscure gathering of stars found halfway between Eta Cygni and 39 Cygni. Have you ever seen it? Few amateurs have, since it is ignored by nearly every star atlas and observing handbook. Roslund 5 consists of some 18 stars ranging in brightness from magnitude 7 to 9 set across $\frac{3}{4}°$. Though best seen in binoculars, it is still a difficult stellar array to pick out among the bevy of surrounding stars.

M29 (NGC 6913) is one of only two Cygnus entries in the famous Messier catalogue. Positioned nearly 2° south of Gamma Cygni at the center of the Northern Cross, this poor open cluster contains about a dozen members scattered over a 7′ area. Studies indicate that the region surrounding M29 contains about 1,000 times the average density of dust found in the Milky Way. Were it not for this obscuring matter, M29 would undoubtedly be a far more impressive sight than it is.

Lynds 906 is the catalogue name for the largest cloud of dark nebulosity found north of the celestial equator. Most of us know the region as the Northern Coalsack. This huge area, just south of the brilliant star Deneb, is easily visible on dark, moonless nights as the northern terminus of the Milky Way's "Great Rift." The Great Rift slices the galactic plane in half as it extends toward constellations south of Cygnus.

Ruprecht 173 is a large, bright open cluster found about one-fourth of the way from Epsilon Cygni to Gamma Cygni. Measuring nearly a degree across, the group is unknown to most amateur observers due to its scattered appearance against the backdrop of the Milky Way. Seven-power binoculars isolate about two dozen stars within Ruprecht 173 that are greater than magnitude 9.5, with the four brightest cluster members set in a narrow diamondlike asterism. Larger glasses are at a disadvantage in viewing Ruprecht 173, since any restriction in field will easily lose the stars' clustering effect.

For variable star observers, Ruprecht 173 hosts **X Cygni**. A well-known Cepheid variable, it is easily followed across its 16-day cycle from a high of 6th magnitude to a low of 7th magnitude.

NGC 6992 is the brightest portion of the Veil Nebula complex and is unquestionably one of the best known sights of the northern summer sky. Believed to be the fragmented remnant of an ancient supernova, the Veil is always a difficult object to see through binoculars and telescopes alike. NGC 6992 measures about a degree in length and appears as a thin crescent of grayish light through 7× and larger glasses. Binoculars are unlikely to reveal much of the delicate texture that is so prominent in photographs.

NGC 7000, seen in Figure 7.11, is a paradoxical object. More commonly called the North American Nebula after its resemblance to the earthly continent, it is known to most amateurs, yet few have actually seen it. Due to the cloud's large expanse (measuring nearly four Full-Moon diameters across), it is often difficult to distinguish against the rich backdrop of the Cygnus Milky Way.

I have always found NGC 7000 much more difficult to spot through a wide-field telescope than with a pair of 7 × 50 binoculars or even with the unaided eye. If one knows what to look for, the North American Nebula is most easily seen without any optical aid at all. A hint for spotting this celestial continent is to look for "Mexico," "Florida," and the "Gulf of Mexico." Collectively, they appear as a slightly brighter, detached portion of the Milky Way just east of Deneb.

IC 5067/70 is nicknamed the Pelican Nebula and is situated off the "east coast" of NGC 7000. Although the Pelican is not prominent directly through binoculars, many observers can detect its outline as contrasted against the large, more easily seen dark nebula **Lynds 935.** Keep in mind that the contrast between IC 5067 and Lynds 935 is very low, so clear, dark skies and clean optics are a must. For those who have tried to find IC 5067 without success, a good rule of thumb is set by neighboring NGC 7000: if the North American Nebula is *easily* visible to the naked eye, then the sky should be clear enough to see the Pelican with giant binoculars.

Harrington 10 is a patch of dark nebulosity found about 7° north-northeast of Deneb. It is easily seen with the naked eye whenever the milky band of our galaxy is visible. Look for a perfectly straight, dark line running perpendicular to the Milky Way. Seven-power glasses serve to heighten its detectability, while 11× and larger glasses begin to show some of the irregularities along the lane's edges that are apparent in wide-field photographs of the region.

61 Cygni is a challenging double star through binoculars. This binary system consists of two brilliant magnitude 5.2 and 6.0 orange type-K suns separated by about 29″ of arc. While 61 Cygni is difficult

Figure 7.11
The North American Nebula, NGC 7000 (left), and the Pelican Nebula, IC 5067 (right), in Cygnus. Photograph by Johnny Horne using an 8-inch Schmidt camera, hypered Technical Pan 2415 film, and a six-minute exposure.

to resolve using only 7×, larger glasses should be capable of just distinguishing the two stars.

Astronomical history was made in 1838. After intensely studying 61 Cygni for years, German astronomer Friedrich Bessel observed that it exhibited a parallax of 0.3″ of arc. Using simple trigonometry, he determined that the distance to the star was approximately 10.3 light years. This was the first time that an estimate of a star's distance had been made relatively accurately. Our modern reckoning shows that 61 Cygni is 11.2 light years distant, a difference of less than 10% from Bessel's original appraisal.

NGC 7039 is a large open cluster set about 2° northwest of the North American Nebula. Through binoculars, seven or eight stars are revealed scattered across the dim glow of unresolved stars and spanning nearly a Moon's diameter.

M39 (NGC 7092), shown in Figure 7.12, is one of the more impressive open clusters found in Cygnus. Famous as a bright, loose congregation of stars located about 10° northeast of Deneb, M39 can be seen with the unaided eye on good nights. Binoculars reveal about two dozen stars ranging from magnitude 7 to 10 set in a broad

Figure 7.12
Open cluster M39 (NGC 7092) in Cygnus. Photographed by George Viscome through a 500-mm lens.

triangular pattern. Against the fainter field stars, the cluster frequently exhibits a striking three-dimensional effect.

IC 5146, better known as the Cocoon Nebula, is a difficult challenge at best for all observers, regardless of equipment, sky clarity, and experience. Even through giant glasses outfitted with contrast-enhancing nebula filters, the Cocoon appears as little more than an amorphous glow surrounding a pair of 9th-magnitude stars. In all, the light of about 20 faint suns illuminate the cloud, but most of these evade detection in binoculars. Deep photographs show the nebula itself to be bisected by dark lanes in a manner similar to the famous Trifid Nebula in Sagittarius. Unfortunately, there is little hope of ever visually spying these features.

Barnard 168 appears as a distinct dark lane apparently extending from the western edge of IC 5146. Through both 7× and 11× glasses, this 10'-wide × 100'-long channel is clearly silhouetted against a multitude of more distant stars. Under dark skies, Barnard 168 and IC 5146 form an intriguing pair.

Delphinus (Del)

Object	Typ	R.A. (2000) h m	Dec. (2000) ° '	Mag.	Size/Sep/ Period	Comments
R	Vr	20 14.9	+09 05	7.6–13.8	284.88 days	Long Period Variable
S 752	**	20 30.2	+19 25	6.6,7.0	106"	288° (1915); 13921
NGC 6934	GC	20 34.2	+07 24	8.9	6'	
Hrr 9	OC	20 38	+13 30			Asterism (Theta Delphini)
U	Vr	20 45.5	+18 05	7.6–8.9p	110 days	Semi-Regular

The constellation of **Delphinus,** the Dolphin, is marked by a kite-shaped asterism of five stars. Though these distant suns are far from the brightest in our sky, the Dolphin's distinctive shape always beckons the attention of naked-eye stargazers. Many amateurs make the mistake of passing over Delphinus, thinking it holds little of interest. Lying to the east of the Milky Way's plane, the Dolphin is indeed noticeably absent of open clusters and vast expanses of nebulosity. However, in their places are some challenging objects awaiting the ardent deep sky observer.

NGC 6934 lies 4° due south of the Dolphin's tail star, Epsilon Delphini, and about ½° north of a pair of 6th-magnitude suns. A 9th-magnitude globular cluster, NGC 6934 is a difficult catch through low-power glasses. Even through 11 × 80 binoculars, it appears as little more than a small glow with but a lone star superimposed on the disk. This star is not a member of NGC 6934, but rather just a chance alignment of two widely separated objects.

Harrington 9 is a pleasant little asterism that surrounds and includes 6th-magnitude Theta Delphini. Theta itself is at the end of a curve of five stars that bends toward the northwest, while just east of Theta is a tight equilateral triangle of 7th-magnitude suns. In all, two dozen 9th-magnitude and brighter stars are found here.

U Delphini is a bright semi-regular variable star that is ideal for study through binoculars. Typically alternating between about visual magnitudes 6 and 8, it has a period of approximately 110 days. Look for this star about 2° north of the diamond's northeastern star, Gamma Delphini, which is also a fine telescopic double star.

Dorado (Dor)

Object	Typ	R.A. (2000) h m	Dec. (2000) ° '	Mag.	Size/Sep/ Period	Comments
NGC 1714	PN	04 52.1	−66 56			In LMC
NGC 1743	PN	04 54.0	−69 12			In LMC
NGC 1755	OC	04 55.0	−68 11	10	2'	In LMC
NGC 1763	DN	04 56.8	−66 24		25'	In LMC
NGC 1818	OC	05 04.2	−66 24	9.8	3'	In LMC
NGC 1850	OC	05 08.5	−68 46	9.3	3'	In LMC
NGC 1866	OC	05 13.5	−65 28	9.8	5'	In LMC
NGC 1910	OC	05 18.1	−69 13		1'	In LMC
S	Vr	05 18.2	−69 15	8.6–11.7p		Irregular; S Dor type; in LMC
NGC 1929	DN	05 22.0	−67 58		20' × 15'	
LMC	Gx	05 23.6	−69 45	0.1	650' × 550'	SBm or Ir Large Magellanic Cloud
NGC 1966	DN	05 26.8	−68 49		13'	In LMC
NGC 1978	GC	05 28.6	−66 14	9.9	3'	In LMC
NGC 2004	OC	05 30.6	−67 17	9.8		In LMC
Beta	Vr	05 33.6	−62 29	3.5–4.1	9.842 days	Cepheid
NGC 2070	DN	05 38.7	−69 06		40'	Tarantula Nebula; in LMC
NGC 2074	DN	05 39.6	−69 27		16' × 10'	In LMC; part of NGC 2070

(Continued on next page)

Dor (continued)

Object	Typ	R.A. (2000) h m	Dec. (2000) ° '	Mag.	Size/Sep/ Period	Comments
NGC 2077	DN	05 40.4	−69 38		15'	In LMC
NGC 2100	OC	05 42.0	−69 14	9.6	2'	In LMC

The far southern constellation of **Dorado**, the Swordfish, was first listed in Bayer's *Uranometria* in 1603. Created from a crooked line of stars that stretches from Caelum in the north to Mensa in the south, Dorado is never well seen north of the equator. This is a great loss to northern hemisphere amateurs and professionals alike, for Dorado plays host to one of the closest galaxies to our own.

The **Large Magellanic Cloud** (Figure 7.13), or Nubecula Major, as it is also known, is a spectacular neighboring galaxy of the Milky Way. It lies a "mere" 163,000 light years from the Milky Way, less than one-tenth the distance to the Andromeda Galaxy. This 50,000-light-year-diameter galaxy is seen as an irregular cloud that spans 11° in our sky. It is easily visible to the unaided eye, even under the light from the Full Moon, with the galaxy's southern boundary extending into adjacent Mensa.

At least 30 billion stars are thought to compose the Large Magellanic Cloud, along with over 400 planetary nebulae, 700 open clusters, 50 diffuse nebulae, and 60 globular clusters.

NGC 1763 is the largest and brightest member of a clump of nebulosity along the northeastern border of the Large Magellanic Cloud. Peering through 7× glasses, a sharp eye might detect a mottled texture to the cloud, as well as the faint shimmer of some of the other nebulous tufts in the area, which include NGC 1760, 1761, 1769, and 1773.

NGC 1910 is the northernmost member of a cluster of 16 open clusters at the center of the Large Magellanic Cloud. Though quite rich, NGC 1910 remains unresolvable in binoculars—that is, except for one notable star.

The star is **S Doradus**, one of the most luminous stars known and the prototype sun for a class of variable stars. Reaching absolute visual magnitude −10 at peak brightness, this young, massive blue star undulates three magnitudes over an irregular period. Analysis of the star's spectrum indicates that these large deviations in magnitude are caused by high-speed ejection of the star's atmosphere. Back on Earth, this is seen as an apparent range from 8th to 11th magnitude,

Figure 7.13
The Large Magellanic Cloud. Note the amazing Tarantula Nebula (NGC 2070) on the galaxy's eastern edge. Photograph by Dennis diCicco of *Sky and Telescope*, using an 8-inch f/1.5 Schmidt camera and ISO 400 film.

which makes S Doradus one of the most distant single stars visible through binoculars.

NGC 2070 is one of the finest diffuse nebulae found anywhere in the sky, yet it is not even in our galaxy! Nicknamed the Tarantula Nebula for its many complex appendages, this glorious cloud defies description. It is so bright that it may even be seen with the unaided eye as a bright patch on the eastern edge of the Large Magellanic Cloud. Binoculars reveal a great network of intertwined rifts swirling about themselves. At the center of NGC 2070, powering the nebula, is a cluster of over 100 supergiant stars.

From its ^{2}⁄$_3$° apparent diameter, astronomers conclude that the Tarantula Nebula must span close to 1,000 light years, making it the largest known emission nebula in the universe. Estimates indicate that if it were at the same distance from Earth as the Orion Nebula (about 1,600 light years), NGC 2070 would span 30°!

Many smaller nebulous knots are faintly visible around the Tarantula: NGC 2048, 2050, 2055, 2074, and 2077 through 2086, among others. All appear as small, dim cloudy puffs.

Draco (Dra)

Object	Typ	R.A. (2000) h m	Dec. (2000) ° '	Mag.	Size/Sep/ Period	Comments
Σ 1516	**	11 15.4	+73 28	7.6,8.1	36"	102° (1940); 8100
NGC 4125	Gx	12 08.1	+65 11	9.8	5' × 3'	E5p
ΟΣΣ 123	**	13 27.1	+64 44	6.7,7.0	69"	147° (1924)
AG	**Vr**	**16 01.7**	**+66 48**	**8.8–11.8p**		**Z And type**
R	**Vr**	**16 32.7**	**+66 45**	**6.7–13.0**	**245.47 days**	**Long Period Variable**
TX	Vr	16 35.0	+60 28	7.9–10.2p	78 days	Semi-Regular
16 + 17	**	**16 36.2**	**+52 55**	**5.4,5.5**	**90"**	**194° (1956); 10129**
AZ	Vr	16 40.7	+72 40	8.0–8.9p		Irregular
AH	Vr	16 48.3	+57 49	8.5–9.3p	158 days	Semi-Regular
AI	Vr	16 56.3	+52 42	7.1–8.1	1.199 days	Eclipsing Binary
Nu	**	**17 32.2**	**+55 11**	**4.9,4.9**	**62"**	**312° (1955); 10628**
Psi	**	17 41.9	+72 09	4.9,6.1	30"	15° (1958); 10759
T	Vr	17 56.4	+58 13	7.2–13.5	421.22 days	Long Period Variable
UW	Vr	17 57.5	+54 40	7.0–8.0		Irregular
NGC 6543	**PN**	**17 58.6**	**+66 38**	**8.8p**	**18"**	
Σ 2278	**	18 02.9	+56 26	7,8,9,10	37",34",201"	26°, 35°, 191° (1949); 11035
39	**	18 23.9	+58 48	5.0,7.4	89"	21° (1956); 11336
Omicron	**	18 51.2	+59 23	4.8,7.8	34"	326° (1949); 11779
UX	Vr	19 21.6	+76 34	5.9–7.1	168 days	Semi-Regular

Winding its way through the northern circumpolar skies of late spring and summer, **Draco**, the Dragon, holds a fine selection of double stars, variable stars, and even a bright planetary nebula for binocular observers. The region is also littered with hundreds of galaxies, but sadly, all are too faint to be easily glimpsed through common binoculars.

AG Draconis is found out in the "boondocks" of northwestern Draco and will probably require some searching before it is found. Look about 5° northwest of Eta Draconis, near a wide triangle of 6th- and 7th-magnitude stars. AG, a Z Andromedae or symbiotic variable, is located just to the triangle's south. Over an irregular period, it may be seen to fluctuate from about visual magnitude 8.0 to 10.5 and may therefore be followed in giant glasses throughout its cycle.

R Draconis, a long period variable, varies like clockwork from magnitude 6.7 to magnitude 13.0 in just under 246 days. Observers should take care not to confuse the variable with a fixed-brightness 9th-magnitude star just to its east.

16 and **17 Draconis** form a wide, easily resolved pair of stars in south-central Draco. Both shine pure white, with 17 the northernmost and slightly brighter of the two. Both also exhibit a common proper motion, which implies that they must be physically associated with each other. Measurements show that 16 orbits 17 in approximately 56 years.

Nu Draconis is another wide double star that can be split in all binoculars. The faintest of the four stars in the Dragon's "head," Nu is composed of two nearly identical 5th-magnitude type-A white stellar jewels. In his classic book *Celestial Objects for Common Telescopes*, the renowned 19th-century deep sky observer Reverend T.W. Webb called these stars "grand," an accurate portrait through modern binoculars as well. Studies indicate that Nu Draconis is a true binary system located nearly 120 light years away. At this distance, the stars' apparent separation of 62 arc-seconds indicates a real separation of about 2,280 astronomical units.

NGC 6543 is one of the sky's ten most distinctive planetary nebulae. Shining with the light of an 8th-magnitude star, it readily reveals its presence in 7× glasses. Although it measures no more than 18″ of arc in diameter, its lucid blue tint, a characteristic shared by many planetaries, separates NGC 6543 from neighboring stars. Larger binoculars will also disclose the nebula's magnitude 9.5 central star, one of few such stars detectable in binoculars. Additionally, a 9th-magnitude field star just to the west joins the nebula to form a pair that might at first be mistaken for a binary star.

Equuleus (Equ)

Object	Typ	R.A. (2000) h　m	Dec. (2000) °　'	Mag.	Size/Sep/ Period	Comments
S	Vr	20 57.2	+05 05	8.0–10.8	3.436 days	Eclipsing Binary
Gamma	**	21 10.3	+10 08	4.7,5.9	353"	153° (1922); 14702

Equuleus, the Colt, is marked by a tiny trapezoid of four faint stars preceding Pegasus in the early autumn sky. With its brightest suns only of 4th magnitude, there is little wonder why most observers do not even know the constellation exists. Occupying an 8° × 10° area, the Colt presents only two binocular objects.

Gamma Equulei marks the northwestern corner of the Equuleus trapezoid. Here, we find two easily resolvable stars spaced nearly 6' of arc apart. Both stars look white to most people, although Gamma[1], a type-F star, may show the slightest hint of yellow.

Eridanus (Eri)

Object	Typ	R.A. (2000) h　m	Dec. (2000) °　'	Mag.	Size/Sep/ Period	Comments
Z	Vr	02 47.9	−12 28	7.0–8.6p	80 days	Semi-Regular
RR	Vr	02 52.2	−08 16	7.4–8.6p	97 days	Semi-Regular
NGC 1232	Gx	03 09.8	−20 35	9.9	7.0' × 5.5'	Sc
NGC 1291	Gx	03 17.3	−41 08	9.0	11'	SBa
NGC 1332	Gx	03 26.3	−21 20	10.3	3' × 1'	E7
T	Vr	03 55.2	−24 02	7.4–13.2	252.24 days	Long Period Variable
β 1042	**	03 58.6	−02 39	7.5,8.5	56"	93° (1913); 2909
W	Vr	04 11.5	−25 08	7.5–14.5	376.63 days	Long Period Variable
h 3628	**	04 12.5	−36 09	7.1,8.0	50"	50° (1933)
NGC 1535	PN	04 14.2	−12 44	9.3p	20" × 17"	
Omicron[2]	**	04 15.2	−07 39	4.4,9.5	83"	104° (1970); 3093
RZ	Vr	04 43.8	−10 41	7.8–8.7	39.282 days	Eclipsing Binary
62	**	04 56.4	−05 10	5.5,9.1	67"	75° (1913)

Flowing across more than 50° of our sky, **Eridanus,** the River, follows a long, narrow path through late autumn's southern sky. One of the oldest constellations, Eridanus has been associated with many waterways, including the Nile of ancient Egypt.

The heavenly river begins near the celestial equator just south of Taurus and winds southward past many smaller star groups, such as Orion and Lepus to its east and Fornax and Phoenix to its west. Along the path, binoculars show several interesting variable and double stars, a lone planetary nebula, and the brightest three of the constellation's many galaxies. Eridanus ends at the brilliant star Achernar (Alpha Eridani). A magnitude 0.53 blue-white giant, Achernar is unknown to most amateurs, as it lies nearly 55° south of the celestial equator.

NGC 1291 is found about 4° southeast of Theta Eridani, near a slim triangle of three 7th-magnitude stars. Through binoculars, this barred spiral galaxy appears as a soft, slightly oval glow featuring a brighter core. Smaller glasses may have a bit of trouble revealing NGC 1291 at first, but giant binoculars should show it plainly.

NGC 1535 is one of the brighter planetaries in the sky, yet few amateurs take the time to find it within the sparse surrounding star field. Shining at 9th magnitude, it appears stellar in binoculars. Only its pale blue color gives it away as a nebula. The 10th-magnitude central star rates a "difficult to impossible" through all binoculars, no doubt because of the cloud's high surface brightness.

Omicron^2 Eridani is a double star that is challenging to resolve in most binoculars. The primary star, Omicron^2 Eridani A, shines at 4th magnitude, while Omicron^2 Eridani B appears as a 9th-magnitude speck to the southeast. What makes this system so interesting is that Omicron^2 B is the brightest white dwarf star seen from Earth. Measurements indicate that it has a mass equal to the Sun, yet this amazing star is believed to be only 17,000 miles across. Nearly 38 billion miles separate these two stellar companions! In addition Omicron^2 has a third stellar component, a red dwarf star believed to have only one-fifth the mass of the Sun.

Fornax (For)

Object	Typ	R.A. (2000) h m	Dec. (2000) ° '	Mag.	Size/Sep/ Period	Comments
R	Vr	02 29.3	−26 06	7.5–13.0	387.85 days	Long Period Variable
NGC 1097	Gx	02 46.3	−30 17	9.3	9' × 6'	SBb
NGC 1316	Gx	03 22.7	−37 12	8.9	4' × 3'	SB0p Radio source
Hrr 2	OC	03 27	−35	6,6,6	30' × 30'	Chi^{1+2+3} asterism
NGC 1360	PN	03 33.3	−25 51		390"	
NGC 1365	Gx	03 33.6	−36 08	9.5	9' × 4'	SB
NGC 1398	Gx	03 38.9	−26 20	9.7	6' × 5'	SBb
S	Vr	03 46.2	−24 24	5.6–8.5		Flared in March, 1899

Carved by Lacaille in 1752 out of stars originally contained in Eridanus, the faint constellation **Fornax,** the Furnace, appears to the naked eye as a nearly starless void in the southern autumn sky. Many faint galaxies populate the region, most belonging to the Fornax Galaxy Cluster. Unfortunately, all but two are too dim for binoculars. There is, however, an interesting stellar asterism in the region, along with a few double and variable stars and a surprising planetary nebula.

NGC 1316 is the brightest galaxy in the Fornax cluster. Confusion over its exact classification has always reigned over this 10th-magnitude galaxy. Some sources list it as a probable S0 spiral, while others refer to it as an elliptical. Whatever its true nature, NGC 1316 remains the target of an intense investigation. Long-exposure photographs reveal loops of material swirling out from it and engulfing NGC 1317, a faint nearby system. Some theorize that NGC 1317 will eventually be absorbed into its neighbor. The phenomenon of a larger galaxy devouring a smaller object is appropriately called *galactic cannibalism* and results in the strong emission of radio noise.

Binoculars reveal NGC 1316 as a fairly large, bright disk of light, highlighted with a brighter central core. Sadly, the features that make this galaxy so unique are visually unobservable due to their subtlety.

Harrington 2 is formed from the stellar triangle Chi^1, Chi^2, and Chi^3 Fornacis, as well as a number of other stars in their immediate vicinity. Chi^1, the westernmost member of the Chi triangle, is an easy optical binary, with the 6th-magnitude secondary star found to the primary's south.

Actually, this asterism appears more like a diamond, thanks to an additional 6th-magnitude star to the Chi triplets' east. In all, nine stars from 5th to 9th magnitude scattered across about ½° compose Harrington 2. Adding to the scene is a semicircle of seven 9th-magnitude stars just to the north of Chi².

NGC 1360 is usually ignored by northern hemisphere observers, as it lies in an empty area of the sky that is difficult to pinpoint. It rewards those who persevere, however, with a faint, though distinct grayish disk of light. Most observers see NGC 1360 as noticeably elliptical, with photographs indicating an apparent diameter of more than 6′ of arc. A faint star may also be seen near the cloud's center through the largest binoculars, although it is not known for certain whether the star is directly associated with the nebula.

NGC 1365 is one of most impressive examples of a barred spiral found south of the celestial equator. Overall, it measures 8′ × 3′ and 10th magnitude. Visual observers will see this galaxy as an oval nebulous patch growing steadily brighter toward its center. The galaxy's characteristic central bar, seen clearly only in photographs, is believed to span some 45,000 light years.

Gemini (Gem)

Object	Typ	R.A. (2000) h m	Dec. (2000) ° ′	Mag.	Size/Sep/ Period	Comments
NGC 2129	OC	06 01.0	+23 18	6.7	7′	
NGC 2158	OC	06 07.5	+24 06	8.6	5′	SW of M35
NGC 2168	OC	06 08.9	+24 20	5.3	28′	M35
BU	Vr	06 12.3	+22 54	5.7–7.5		Irregular
Eta	Vr	06 14.9	+22 30	3.2–3.9	232.9 days	Semi-Regular
Cr 89	OC	06 18.0	+23 38	5.7p	35′	
Nu	✳✳	06 29.0	+20 13	4.2,8.7	113″	329° (1924); 5103
X	Vr	06 47.1	+30 17	7.5–13.6	263.72 days	Long Period Variable
Zeta	✳✳	07 04.1	+20 34	3.8,8.0	96″	350° (1925); 5742; A = var (3.6–4.2)
NGC 2331	OC	07 07.2	+27 21	9p	18′	
R	Vr	07 07.4	+22 42	6.0–14.0	369.81 days	Long Period Variable
NGC 2355–6	OC	07 16.9	+13 47	10p	9′	
V	Vr	07 23.2	+13 06	7.8–14.9	275.07 days	Long Period Variable
Σ 1090	✳✳	07 26.5	+18 31	7.3,8.2	61″	97° (1921); 6073
NGC 2395	OC	07 27.1	+13 35	8.0	12′	
NGC 2392	PN	07 29.2	+20 55	8.3	13″	Eskimo Nebula
Alpha	✳✳	07 34.6	+31 53	1.9,8.8	73″	164° (1955); 6175; Castor

(Continued on next page)

Gem (continued)

Object	Typ	R.A. (2000) h m	Dec. (2000) ° '	Mag.	Size/Sep/ Period	Comments
NGC 2420	OC	07 38.5	+21 34	8.3	10'	
U	Vr	07 55.1	+22 00	8.2–14.9	103 days?	Irregular; U Gem prototype

Apparently standing on the northern perimeter of the Milky Way is **Gemini,** the Twins. Marked by the bright stars Castor (an interesting binary star itself) and Pollux, Gemini is one of the best known of the winter constellations. Within its borders for observers with binoculars are many fine examples of double and variable stars, open clusters, and even a planetary nebula.

M35 (NGC 2168), shown in Figure 7.14, is one of the premier open clusters of the northern winter sky. Located at the foot of the twin Castor, M35 is a marvelous sight in binoculars and may even be seen with the unaided eye on crystal clear nights. Low-power glasses reveal

Figure 7.14
The fine open cluster M35 (NGC 2168) in Gemini. Note the rich cluster NGC 2158 on the lower right-hand edge of M35. George Viscome made this 12-minute exposure on ISO 400 film through a 500-mm lens.

the brightest half-dozen or so cluster stars against the strong glow from an additional 200 fainter orbs that constitute this magnificent stellar gathering. These brightest members are brilliant indeed and average about 400 times the luminosity of our Sun. Estimates place M35 at about 2,800 light years away and some 40 light years across.

Just southwest of M35 is the very rich and very distant open cluster **NGC 2158**. Although it is listed between magnitude 8 and 9 in many observing handbooks (including this one), I have always felt this to be a bit optimistic. Try as I might, I have never been able to see NGC 2158 in 7×50 binoculars, although it *should* be visible if it is actually that bright. Instead, I find it requires at least 11 × 80's to be seen, and even then, it is seen as a dim glow against a striking star field. When we look toward NGC 2158, we are seeing an object nearly 16,000 light years away, close to the outer fringes of our galaxy.

BU Geminorum is an intriguing irregular variable star that is well suited for study through binoculars. This red sun wanders between magnitudes 5.7 and 7.5 across an unpredictable time interval. Interestingly, some believe that BU Geminorum is actually a binary star and may exhibit a 32-year cycle. Further study of the star's behavior is required before this question can be answered.

Collinder 89 resides amidst a line of stars located about midway between M35 and Mu Geminorum. Studies identify 15 stars within this open cluster, although only about half a dozen can be seen clearly in binoculars. Marking the boundaries of Collinder 89 are 6th-magnitude 9 Geminorum on its western edge and 10 Geminorum at the eastern limit. While these suns help point the way to this rather obscure open cluster, the lack of any stellar concentration still makes it a difficult object to identify positively.

NGC 2392, popularly known as the Eskimo Nebula, is one of the brightest planetary nebulae in the winter sky. It shines at 8th magnitude and spans nearly 13″ of arc across, a winning combination for binoculars. Actually, through low-power glasses, NGC 2392 can be mistaken for the southern half of a double star, as it is found just south of an unrelated 9th-magnitude field star. What gives the planetary away is its tiny, vibrant blue disk, which is far more colorful than any of the surrounding stars. Larger glasses will also reveal the nebula's central star, which blazes away at 10th magnitude.

Photographs of NGC 2392 reveal two rings of nebulosity separated by a dark zone. The inner ring has a mottled appearance that loosely resembles a human face, while the dim outer ring encircles it like the fur-lined hood of a jacket—ergo, the name "Eskimo" Nebula.

U Geminorum is the prototype sun of one of the most fascinating classes of variable stars known. Normally, this cataclysmic variable lies at about 15th magnitude. Every 100 days or so, however, its brightness rockets up to between magnitudes 8 and 9 in a period of less than one day. Astronomers believe that these amazing novalike flares are the result of interplay between a red dwarf paired with a white dwarf and encircled by a ring of cascading gaseous material.

Grus (Gru)

Object	Typ	R.A. (2000) h m	Dec. (2000) ° '	Mag.	Size/Sep/ Period	Comments
BrsO 15	**	21 48.3	−47 18	5.7,8.7	55″	356° (1931)
R	Vr	21 48.5	−46 55	7.4–14.9	331.89 days	Long Period Variable
Pi¹	Vr	22 22.7	−45 57	5.4–6.7	150 days	Semi-Regular Red!
Pi¹⁺²	**	22 23.1	−45 56	6.6,5.8	365″	Optical
T	Vr	22 25.7	−37 34	7.8–12.3p	136.53 days	Long Period Variable
S	Vr	22 26.1	−48 26	6.0–15.0	401.37 days	Long Period Variable
Delta²	**	22 29.8	−43 45	4.1,9.0	61″	212° (1953)

Situated just south of the bright autumn star Fomalhaut is the small constellation **Grus,** the Crane. Introduced in 1603 by Bayer, Grus' three brightest stars form an almost-right triangle, with a string of fainter stars set along its hypotenuse. Since the constellation is far from the galactic plane, there are neither clusters nor nebulae to attract our attention, and all of the galaxies within are too faint for binocular detection. Still, the Crane contains several interesting double and variable stars that are visible in binoculars.

Pi¹⁺² Gruis combine to form a fine optical binary star. The former is a striking orange-red semi-regular variable star that may be seen to fluctuate between magnitudes 5.4 and 6.7 over 150 days. The latter glistens with a nicely contrasting yellowish glint and shines steadily at magnitude 6.6. With over 4′ of arc between them, even the lowest power pair of binoculars will have no trouble detecting both stars. While they are not actually related to each other, both Pi¹ and Pi² are true binary systems. Sadly, their respective companion suns require telescopes to be resolved.

Delta[1+2] **Gruis** form another wide, unrelated pair of stars. Found in central Grus, both of these 4th-magnitude suns may be spotted with the unaided eye and are easily seen through 6× glasses. Seven-power and larger binoculars will also faintly reveal the real 9th-magnitude companion of Delta[2], set about 1′ of arc away. Care must be taken to see Delta[2] B successfully, as the glare from Delta[2] A and Delta[1] can easily overwhelm an observer's eyes.

Hercules (Her)

Object	Typ	R.A. (2000) h m	Dec. (2000) ° ′	Mag.	Size/Sep/ Period	Comments
ST	Vr	15 50.8	+48 29	8.8–10.3p	148.0 days	Semi-Regular
X	Vr	16 02.7	+47 14	7.5–8.6p	95.0 days	Semi-Regular
SX	Vr	16 07.5	+24 55	8.6–10.9p	102.90 days	Semi-Regular
RU	Vr	16 10.2	+25 04	6.8–14.3	485.49 days	Long Period Variable
Hrr 7	**OC**	**16 18**	**+13**		**100′ × 15′**	**Asterism**
U	Vr	16 25.8	+18 54	6.5–13.4	406.05 days	Long Period Variable
g	Vr	16 28.6	+41 53	5.7–7.2p	70 days	Semi-Regular 30 Herculis
37	**	16 40.6	+04 13	5.8,7.0	70″	230° (1932); 10149
NGC 6205	**GC**	**16 41.7**	**+36 28**	**5.9**	**16′**	**M13**
NGC 6210	**PN**	**16 44.5**	**+23 49**	**9.3p**	**>14″**	
NGC 6229	GC	16 47.0	+47 32	9.4	5′	
S	Vr	16 51.9	+14 56	6.4–13.8	307.44 days	Long Period Variable
UW	Vr	17 14.4	+36 22	8.6–9.5p	100 days	Semi-Regular
Alpha	**Vr**	**17 14.6**	**+14 23**	**3.1–3.9**	**90 days**	**Semi-Regular**
NGC 6341	**GC**	**17 17.1**	**+43 08**	**6.5**	**11′**	**M92**
RS	Vr	17 21.7	+22 55	7.0–13.0	219.65 days	Long Period Variable
Z	Vr	17 58.1	+15 08	7.3–8.1p	3.993 days	Eclipsing Binary
DoDz 9	**OC**	**18 08.8**	**+31 32**		**34′**	
T	Vr	18 09.1	+31 01	6.8–13.9	165.01 days	Long Period Variable
AC	Vr	18 30.3	+21 52	7.4–9.7p	75.462 days	RV Tauri type

Passing nearly overhead during the summer months in mid-northern latitudes, **Hercules,** the Giant, is a rather difficult constellation to trace with the unaided eye. Its brightest stars are only of the 3rd magnitude and appear rather randomly patterned. Only the four stars in the Hercules Keystone can be quickly identified. Hercules holds some spectacular deep sky objects, including a pair of magnificent globular clusters, a bright planetary nebula, and a host of others.

Harrington 7 might be called the Zigzag Cluster (though it is only an asterism), as it is composed of 17 magnitude 7 to magnitude 9 stars winding across a little over 1½° of sky. Using giant glasses, look for this faint group about 6° directly south of Gamma Herculis.

M13 (NGC 6205) is considered by most amateurs to be the premier globular cluster in northern skies. Shining at 6th magnitude, it may be seen with the naked eye as a dim blur one-third of the way from Eta Herculis to Zeta Herculis along the western edge of the Keystone. Its fuzzy globe measures 16′ of arc across and is set between a pair of 7th-magnitude field stars.

The discovery of the Great Hercules Cluster is credited to Halley, who first spotted it in 1714. A half-century later, Messier described it as merely a starless nebula. While Messier's description does not accurately portray the globular's true nature, it does fit the modern view through low-power binoculars. Not until we view M13 through 20× and larger glasses do a few of the hundreds of thousands of stars that belong to this 23,000-light-year-distant object come through.

NGC 6210 is bright enough to be seen through most binoculars, but the small size of this planetary nebula can make it difficult to distinguish from the surrounding stars. On the other hand, the nebula's characteristic bluish color enables it to stand out in a crowded star field. Unfortunately, the magnitude 12.5 central star remains a telescopic target only.

Alpha Herculis, properly named Rasalgethi, was first discovered to be a variable star in 1795 by William Herschel. An ideal semi-regular star for study through binoculars, this red giant rallies from a low of magnitude 3.9 to a peak of magnitude 3.1 over a widely varying period that averages 90 days.

Alpha is also an attractive telescopic double star, with the variable Alpha A accompanied by a magnitude 5.4 blue-green companion. Unfortunately, while this is a striking binary star in telescopes, it is impossible to resolve in binoculars, as Alpha B lies less than 5″ away from Alpha A.

M92 (NGC 6341) is Hercules' oft-forgotten second Messier object. Were it not for its more glamorous neighbor to the south, M92 would undoubtedly be a better known showpiece. Discovered by Bode in 1777, M92 is easily located about 6° north of Pi Herculis, the northwest corner star in the Keystone. It is unlikely that common binoculars will show M92 as more than a misty glow; partial resolution of its approximately 100,000 stars is reserved for 4-inch and larger telescopes. Wide-angle glasses just squeeze M92 and M13 into the same 10° field, permitting the observer to make a direct comparison between the two.

Dolidze-Dzimselejsvili 9 is a large, little-known open cluster found in southeastern Hercules less than a degree west of the star 104 Herculis. Binoculars reveal about two-thirds of this group's 15 stellar affiliates loosely scattered across half a degree of sky.

Horologium (Hor)

Object	Typ	R.A. (2000) h m	Dec. (2000) ° '	Mag.	Size/Sep/ Period	Comments
Δ 7	**	02 39.7	−59 34	7.2,7.5	37″	96° (1953)
R	Vr	02 53.9	−49 53	4.7–14.3	403.97 days	Long Period Variable
NGC 1261	GC	03 12.3	−55 13	8.4	7′	

Horologium, the Clock, is a long, thin constellation that runs parallel to Eridanus in the late-autumn southern sky. First contrived by Lacaille in 1752, this stellar timepiece is usually missed by naked eye skywatchers unless they have a dark, unobstructed view and sharp eyes. Even with the aid of binoculars, the selection of viewable objects is limited.

R Horologii is a fine long period variable for study with binoculars. When near maximum brightness, an event occurring every 404 days, R may even be seen with the unaided eye. At minimum, however, this reddish sun drops nearly ten full magnitudes, which places it well out of binocular range. Look for R about 10° south of Theta Eridani and about 1° north of an easily-split optical binary star.

NGC 1261, an 8th-magnitude globular cluster, lies some 6° southeast of R Horologii. There, binoculars find a small, ghostly ball of celestial cotton, condensing to a brighter core. A faint foreground star close to its northeastern edge will help observers zero in on this 44,000-light-year-distant swarm.

Hydra (Hya)

Object	Typ	R.A. (2000) h m	Dec. (2000) ° '	Mag.	Size/Sep/ Period	Comments
NGC 2548	OC	08 13.8	−05 48	5.8	55'	M48
RT	Vr	08 29.7	−06 19	7.0–11.0	253 days	Semi-Regular
h 99	**.	08 37.8	−06 48	6.8,9.1	61"	202° (1918); 6900
RV	Vr	08 39.7	−09 35	8.7–10.0p	116 days	Semi-Regular
27	**	09 20.5	−09 33	5.0,6.9	229"	211° (1923); 7311
Y	Vr	09 51.1	−23 01	8.3–12.0p	302.8 days	Semi-Regular
NGC 3242	PN	10 24.8	−18 38	8.6p	16"	Ghost of Jupiter
U	Vr	10 37.6	−13 23	7.0–9.2p	450 days	Semi-Regular
FF	Vr	10 37.9	−12 01	8.2–10.3p	85 days	Semi-Regular
V	Vr	10 51.6	−21 15	6.5–12	533 days	Semi-Regular Carbon star
TT	Vr	11 13.2	−26 28	7.5–9.5p	6.953 days	Eclipsing Binary
h 4465	**	11 41.7	−32 30	5.3,8.4	67"	44° (1919)
NGC 4590	GC	12 39.5	−26 45	8.2	12'	M68
R	Vr	13 29.7	−23 17	3.0–11.0	389.6 days	Long Period Variable
NGC 5236	Gx	13 37.0	−29 52	7.6	11' × 10'	Sc M83
W	Vr	13 49.0	−28 22	7.7–11.6p	397.4 days	Semi-Regular
RU	Vr	14 11.6	−28 53	7.2–14.3	333.19 days	Long Period Variable

Of the 88 constellations depicted in the sky, none spans so great a length as **Hydra,** the Water Serpent. Westernmost Hydra, lying just east of Canis Minor and winter's Milky Way, consists of a four-star trapezoid marking one of the serpent's seven heads. The long, slithering body of the Hydra winds its way across the southern spring sky to end near Libra, Scorpius, and the gateway to the summer Milky Way.

Although its stars are faint and at times rather hard to follow, the great length of Hydra contains many fine examples of nearly every type of deep sky object. It is a truly universal constellation!

M48 (**NGC 2548**) is found in a seemingly empty part of western Hydra, about 14° southeast of the bright winter star Procyon. Though considered today to be one and the same as open cluster NGC 2548, M48 had been classified for years as a missing Messier object. The confusion stemmed from Messier's own recollection of its position in the sky. When he discovered it in 1771, Messier erroneously placed this stellar swarm about 4° farther north than it actually is. While there is no cluster to be found at his specified location, Messier's description agrees very closely with that of NGC 2548.

Due to their wider fields of view, binoculars win over telescopes with "normal" focal lengths as the instruments of choice for viewing the full-degree spread of M48. Over a dozen of this cluster's 80 stars may be seen through 7× and larger glasses, with an especially noticeable, tight triangular asterism at its center. The remaining cluster stars blend into a homogeneous glow that stands out well against the surrounding sky.

27 Hydrae is a wide, easily split double star about 2° southwest of Alphard (Alpha Hydrae), the brightest star in the constellation. The 5th-magnitude, spectral type-G primary star exhibits the species' characteristic yellowish tint, especially if its image is just nudged out of focus. The accompanying 7th-magnitude secondary sun, separated from the primary by nearly 4' of arc, shines pure white and is easily visible even with the lowest power opera glass. The pair has an observed common proper motion, indicative of a true binary.

NGC 3242 is the brightest planetary nebula visible in the spring sky from the northern hemisphere. Found about 2° south of 4th-magnitude Mu Hydrae, NGC 3242 will easily show its tiny 8th-magnitude blue disk in 7 × 50 binoculars. A pronounced bulge is evident through higher power glasses, but the 11th-magnitude central star will probably remain out of range. This 2,600-light-year-distant planetary has been nicknamed the Ghost of Jupiter for its similar appearance to our largest planet. Large telescopes and photographs, however, show a strong resemblance to a human eye, with a fainter outer shell of gas surrounding the brighter core.

M68 (NGC 4590) is a relatively small, faint globular cluster found about 4° south of the Corvus trapezoid. Assessed at 8th magnitude, it remains only a dim ball of fluff when viewed through binoculars. Modern estimates place M68 at just over 30,000 light years away and approaching us at about 117 kilometers per second (78 miles per second).

R Hydrae is one of the easiest long period variable stars for amateur astronomers to observe. Sharp-eyed observers have been known to spot it without any optical aid at all as it rises to 4th magnitude or even brighter every 390 days. Binoculars display it as a dazzling red beacon when near its peak. Cycling downward, R bottoms out at 11th magnitude, which still might be within range of large binoculars if one knows exactly where to look.

Since the discovery of R Hydrae in 1704, its period has shortened from 500 days to the present 390. Astronomers are not sure what is going on within this red giant star, but it certainly warrants close scrutiny in the future.

M83 (NGC 5236) is a frequently neglected showpiece object of the southern spring sky. It was discovered in 1752 by Lacaille, who saw it as only an ill-defined nebula. Photographs reveal its true nature as a classic Sc spiral galaxy with three widespread arms.

Riding the Hydra-Centaurus border, M83 is bright enough to be seen through nearly all binoculars. Look for a bright stellar nucleus engulfed in the soft glow of the spiral arms. Set among an attractive star field, M83 is sure to become one of your favorite seasonal targets.

Hydrus (Hyi)

Object	Typ	R.A. (2000) h m	Dec. (2000) ° '	Mag.	Size/Sep/ Period	Comments
NGC 602	DN	01 29.6	−73 33			In SMC?
Pi^{1+2}	**	02 14.2	−67 50	5.6,5.7	900″	Optical

The far southern constellation **Hydrus** was first depicted by Bayer in 1603. Representing a Water Snake, its stars wind their way southward from near Archernar in Eridanus toward the south celestial pole. Far from the Milky Way plane, this constellation reveals but a lone double star and a single nebula through binoculars.

NGC 602 is a small patch of diffuse nebulosity that seems curiously out of place in this constellation so far from the plane of our galaxy. However, with the Small Magellanic Cloud only a few degrees away in neighboring Tucana, many astronomers believe that NGC 602 may actually belong to that galactic system. Dark, clear skies are needed to reveal this nebula through binoculars. Large glasses show it only as a tiny, oval smudge that may require averted vision to be seen at all. Photographs reveal NGC 602 to be sprinkled with faint stars and divided in half by a dark rift.

Pi^{1+2} Hydri lies about halfway between Alpha and Gamma Hydri. Separated by 15′ of arc, this pair of 6th-magnitude stars is an easy target for all binoculars. Color-sensitive observers will also immediately notice that both of these distant suns shine with a red-orange tint.

Indus (Ind)

Object	Typ	R.A. (2000) h m	Dec. (2000) ° '	Mag.	Size/Sep/ Period	Comments
S	Vr	20 56.4	−54 19	7.9–17.0p	399.95 days	Long Period Variable
T	Vr	21 20.2	−45 01	7.7–9.4p	320 days	Semi-Regular

Indus, the Indian, was one of several new constellations that first appeared in Bayer's *Uranometria*. With borders stretching from Microscopium in the north to Octans in the south, Indus bridges a large gap in the far southern autumn sky. Unfortunately, while populated with many faint galaxies, the region holds only a pair of variable stars for the observer with binoculars.

S Indi may be found about 1½° northwest of 5th-magnitude Mu Indi. Over the course of approximately 13 months, S varies between a maximum visual magnitude of about 6.5 and a minimum magnitude of 15.5. Many observers should detect the characteristic reddish color of this long period variable when it is near its brightest light.

T Indi rides close to the Indian's extreme northern border. Varying between about visual magnitudes 6.5 and 8.5, it remains visible through most glasses over its entire 320-day cycle and is therefore ideal for systematic study by southern observers.

Lacerta (Lac)

Object	Typ	R.A. (2000) h m	Dec. (2000) ° '	Mag.	Size/Sep/ Period	Comments
NGC 7209	OC	22 05.2	+46 30	6.7	25'	
NGC 7243	OC	22 15.3	+49 53	6.4	21'	
S	Vr	22 29.0	+40 19	7.6–13.9	241.80 days	Long Period Variable

Hevelius devised the constellation **Lacerta,** the Lizard, in 1687 from a set of faint stars wedged between Cygnus and Pegasus. Although the plane of the Milky Way skims through Lacerta's northern border, we find only a pair of open clusters and a single variable star suitable for viewing with binoculars.

NGC 7209 lies less than a degree in from the Cygnus border and is set in a rich field of stars. The brightest suns of this open cluster just break 9th magnitude, while most of its 25 members shine between 10th and 11th magnitude. Through most glasses, this translates the cluster into a nebulous glow with only a couple of stellar points set within.

NGC 7243 is a bit more satisfying than NGC 7209 in binoculars. Floating among a fine Milky Way field, this cluster holds 40 stars, of which eight are bright enough to be seen in glasses. Once again, the remaining cluster suns pool their light into a cloudy, circular glow.

Leo (Leo)

Object	Typ	R.A. (2000) h m	Dec. (2000) ° '	Mag.	Size/Sep/ Period	Comments
NGC 2903	Gx	09 32.2	+21 30	9.0	13' × 7'	Sb+
7	**	09 35.9	+14 23	6.2,10.0	41"	80° (1946); 7448
R	Vr	09 47.6	+11 26	4.4–11.3	312.4 days	Long Period Variable
Alpha	**	10 08.4	+11 58	1.4,7.7	177"	307° (1924); 7654; Regulus
NGC 3351	Gx	10 44.0	+11 42	9.7	7' × 5'	SBb M95
NGC 3368	Gx	10 46.8	+11 49	9.2	7' × 5'	Sbp M96
NGC 3379	Gx	10 47.8	+12 35	9.3	5' × 4'	E1 M105
NGC 3521	Gx	11 05.8	−00 02	8.9	10' × 5'	Sb+
NGC 3623	Gx	11 18.9	+13 05	9.3	10' × 3'	Sb M65
NGC 3627	Gx	11 20.2	+12 59	9.0	9' × 4'	Sb+ M66
NGC 3628	Gx	11 20.3	+13 36	9.5	15' × 4'	Sb
Tau	**	11 27.9	+02 51	5.1,8.0	91"	176° (1932)

After perhaps only Ursa Major, **Leo,** the Lion, is the most easily recognizable constellation of the northern spring sky. The celestial king of the beasts is marked by a distinctive curve of stars resembling a backwards question mark or a farmer's sickle and a large triangle of stars to its east. Leo is well known to deep sky observers for its many galaxies, both dim and bright. Eight are visible in binoculars under skies free of interference from light pollution.

NGC 2903 is a large, relatively faint spiral galaxy found to the west of Epsilon Leonis at the tip of Leo's sickle. Binoculars display it as a nebulous 5' × 3' patch highlighted by a brighter central core. Long-exposure photographs more than double the galaxy's extent, as well as reveal the jumbled texture of its spiral arms.

R Leonis is spring's selected variable star for those viewing through binoculars (a finder chart is found in Chapter 6). Its location is easily pinpointed about 6° due west of Regulus and just to the south of 19 Leonis. Over the course of 312 days, this long period variable is seen to fluctuate between about magnitudes 4.4 and 11.3. Thanks to this broad magnitude range and the strong reddish glow of R Leonis, it is a seasonal favorite of many.

Alpha Leonis, known to most as Regulus, marks the handle of the Lion's sickle. Copernicus is credited with dubbing this star Regulus, meaning "Little King" in Latin, although he was not the first to refer to it as kinglike. Many ancient cultures, including the Arabians, the Babylonians, and the Akkadians of Mesopotamia, also viewed this star as celestial royalty.

Accompanying Regulus is a small 8th-magnitude star that is easy to spot in binoculars. It may be found nearly 3' of arc from the brilliant primary sun. A third companion of Regulus has also been detected, but its visibility is restricted to large telescopes. There is little doubt that these comprise a true multiple system, as all three stars exhibit the same proper motion.

M95 (NGC 3351), M96 (NGC 3368), and **M105 (NGC 3379)** form a triangle of faint galaxies in central Leo. All three, along with several even fainter surrounding galaxies, are believed to be about 29 million light years from the Milky Way.

M96 is the brightest of the three galaxies. Classified as an Sb spiral, M96 is seen as a slightly oval glow that brightens toward the center. M95, a barred spiral, is smaller and about half a magnitude fainter. When we view M95 and M96, we are seeing only their brighter central cores; the surrounding spiral arms are too faint for detection through binoculars. Scanning a scant ¾° to the north of M95 and M96 brings M105 into view. This elliptical galaxy shows little more than a faint oval disk devoid of any central nucleus.

M65 (NGC 3623), M66 (NGC 3627), and **NGC 3628** (all shown in Figure 7.15) may be found below the triangular hindquarters of Leo. Studies indicate that all three lie about 31.6 million light years away and appear to belong to the Leo Galaxy Group, just like M95, M96, and M105.

M66 is the easiest of the three to pick out. A highly-inclined Sb spiral, its wide elliptical disk and brighter oval center can be clearly

Figure 7.15
A fine galactic trio in Leo: NGC 3628, M65 (NGC 3623), and M66 (NGC 3627) clockwise from the top. Photograph by Martin Germano through an 8-inch f/5 Newtonian reflector using hypered Technical Pan 2415 film and a 45-minute exposure.

made out in 10× glasses. Although slightly fainter, M65 is nearly identical to M66 in both visual appearance and form. Both were discovered in 1780 by Méchain. It is ironic to note that seven years earlier, on November 1, 1773, a comet discovered by Messier passed between these two galaxies, yet Messier failed to detect them. It has been suggested that Messier's failure might have been due to the comet's brightness.

Neither Messier nor Méchain spotted a third galaxy to the pair's north. NGC 3628 is larger and fainter than both M65 and M66. Photographs show it to be an edge-on spiral highlighted by a thin lane of obscuring nebulosity. Unless viewed under superb conditions, the visibility of this galaxy will be restricted to giant binoculars.

Leo Minor (LMi)

Object	Typ	R.A. (2000) h m	Dec. (2000) ° '	Mag.	Size/Sep/ Period	Comments
R	Vr	09 45.6	+34 31	6.3–13.2	371.93 days	Long Period Variable
Hrr 6	OC	10 10	+31 30		45'	Sailboat Cluster

Leo Minor occupies a small, rather desolate region of the northern spring sky between Ursa Major and Leo. Although it is strewn with many galaxies, all are far below the reach of most glasses. Indeed, for observers using binoculars, within Leo Minor are a lone variable star and a unique asterism.

R Leonis Minoris is a fine variable star to follow through binoculars. Over the course of 372 days, this crimson sun fluctuates between a maximum magnitude of 6.3 to a minimum light of magnitude 13.2. Therefore, observers viewing through giant glasses should be able to monitor the star's activity across about three-fourths of its cycle. Unfortunately, finding R Leonis Minoris is much easier said than done. It may be spotted about 10° north of the sickle of Leo and about 5° east of 3rd-magnitude Alpha Lyncis.

Harrington 6 was first brought to the world's attention in *Deep Sky* magazine (No. 23, Summer 1988) by Daniel Hudak of Youngstown, Ohio. He dubbed it the celestial "Sailboat" for its unique shape. Through binoculars, the Sailboat seems to have capsized, as its mast points toward the south. The ship's bow is marked by 7th-magnitude 22 Leonis Minoris, the group's brightest star. Eight additional stars complete the asterism.

Lepus (Lep)

Object	Typ	R.A. (2000) h m	Dec. (2000) ° '	Mag.	Size/Sep/ Period	Comments
β 314	**	04 59.0	−16 23	5.9,8.2	53"	34° (1914); 3588
R	Vr	04 59.6	−14 48	5.5–11.7	432.13 days	Long Period Variable; Red!
T	Vr	05 04.8	−21 54	7.4–13.5	368.13 days	Long Period Variable
RX	Vr	05 11.4	−11 51	5.0–7.0		Irregular
S 476	**	05 19.3	−18 31	6.2,6.4	39"	18° (1952); 3910
NGC 1904	GC	05 24.5	−24 33	8.4	3'	M79
h 3780	**	05 39.3	−17 51	6,9,8,8	89",76",129"	136°, 7°, 299° (1916); 4254

(Continued on next page)

Lep (continued)

Object	Typ	R.A. (2000) h m	Dec. (2000) ° ′	Mag.	Size/Sep/ Period	Comments
NGC 2017	OC	05 39.4	−17 51			
Gamma	**	05 44.5	−22 27	3.7,6.3	96″	350° (1957); 4334
S	Vr	06 05.8	−24 12	7.1–8.9p	90 days	Semi-Regular

Lepus, the Hare, bounds across the southern winter sky directly beneath the conspicuous constellation Orion. Perhaps because of its much more prominent neighbor to the north, amateurs tend to ignore the many fine objects in Lepus. For those willing to delve a little deeper, the Hare holds many fine deep sky "carrots."

R Leporis, a classic long period star for observers with binoculars, is our featured winter variable (a finder chart is in Chapter 6). Radiating a magnificent red hue, R varies between magnitudes 5.5 and 11.7 across a 432-day cycle. Discovered in 1845 by the English astronomer J.R. Hind, the star displays such an intense ruddy color when near maximum that it has been christened the "Crimson Star."

Spectral analysis indicates R Leporis to be a rare *carbon star*. Carbon stars are low-temperature red giants whose spectra reveal the strong presence of carbon compounds. Other noted carbon stars visible in binoculars are V Hydrae and S Cephei.

M79 (NGC 1904) is the only winter globular cluster visible through northern hemisphere binoculars. Méchain was first to spot this 8th-magnitude object in October 1845. He recorded it as simply a nebulous object, an apt description of its appearance through binoculars as well since resolution is not possible in anything less than moderately large amateur instruments. M79 may most easily be found by drawing an imaginary line from Alpha Leporis to Beta Leporis and continuing it an equal distance to the latter's south.

NGC 2017 is a small, weak clump of stars set about 1½° east of Alpha Leporis. Binoculars reveal only four suns of 6th to 9th magnitude, with colors ranging from yellow and orange to blue. All belong to the multiple star **h 3780** and are resolvable in 7× glasses.

Gamma Leporis is a splendid double star awash in vivid color. The system's 3rd-magnitude primary sun is a striking yellow orb, while the 6th-magnitude secondary star appears as an orangish point of light. Separated by 96″ of arc, they form one of the finest binary stars in the binocular sky.

Libra (Lib)

Object	Typ	R.A. (2000) h m	Dec. (2000) ° '	Mag.	Size/Sep/ Period	Comments
SHJ 179	**	14 25.5	−19 58	6.4,7.6	35"	296° (1955); 9258
Alpha[1+2]	**	14 50.9	−16 02	2.8,5.2	231"	314° (1913)
Delta	Vr	15 01.0	−08 31	4.9−5.9	2.327 days	Eclipsing Binary
Iota	**	15 12.2	−19 47	5.1,9.4	58"	111° (1919); 9532
SHJ 195	**	15 14.5	−18 26	7.1,8.1	47"	140° (1916)
NGC 5897	GC	15 17.4	−21 01	8.6	13'	
RS	Vr	15 24.3	−22 55	7.0−13.0	217.65 days	Long Period Variable
RU	Vr	15 33.3	−15 20	7.2−14.2	316.56 days	Long Period Variable

Libra, the Balance Scales, occupies the dim space along the ecliptic between the brighter constellations Virgo and Scorpius. Libra is a relatively recent addition to the zodiac. In his *Almagest* of 150 A.D., Ptolemy assigned its stars to the claws of Scorpius. This origin is still evident today by the names of Libra's two brightest stars, Zuben El Genubi and Zuben Eschamali, which are Arabic for "north claw" and "south claw," respectively. It was not until the reign of Julius Caesar that a separate identity was given to the stars of Libra.

With the naked eye, we can just make out the diamond pattern of 3rd- and 4th-magnitude stars that form the main figure of Libra. Apart from a faint globular cluster and a few double and variable stars, there are few objects within its borders of interest to amateur observers.

Alpha[1+2] Librae, distinctively named Zuben El Genubi, is a widely separated binary star that may easily be resolved with only the slightest optical aid. Alpha[1], also a spectroscopic binary, shines at magnitude 2.8, while Alpha[2] is magnitude 5.2. Nearly 4' of arc separate the two in our sky. Studies indicate that they form a true physical pair and that they lie about 65 light years away. If this is indeed the case, then the stars of Alpha[1+2] Librae are separated by about 48,000 astronomical units.

Iota Librae proves to be a challenging binary in binoculars, not so much because of the stars' closeness, but because of their disparity in brightness. The 5th-magnitude primary sun is paired with an elusive magnitude 9.4 companion separated by nearly a minute of arc to its southeast. Since their positions were first measured in 1782 by William Herschel, both stars have demonstrated a common proper motion, indicating that they form a true physical pair. In addition, as recently as 1940, Iota Librae A was detected to have a companion nearly equal in magnitude to it. At its discovery, the magnitude 5.8 B star was separated from the A star by only 0.2" of arc. Iota Librae C, as the 9th-magnitude component is now known, has a companion of equal magnitude to itself, with about 2" of arc between the stars.

NGC 5897 is a difficult globular cluster to see through even the largest glasses. Look for its ill-defined smudge of grayish light about $1\frac{1}{2}°$ to the southeast of Iota Librae. Perhaps aiding in detection is an arc of three 8th-magnitude stars just to the cluster's northwest. Owing to the cluster's unusually weak concentration of stars, it appears only about half as large as the photographically recorded diameter of 13' implies. Estimates place NGC 5897 at a little over 39,000 light years from the Solar System.

Lupus (Lup)

Object	Typ	R.A. (2000) h m	Dec. (2000) ° '	Mag.	Size/Sep/ Period	Comments
NGC 5593	OC	14 25.8	−54 49	9.0	8'	
Tau[1+2]	**	14 26.1	−45 13	4.6,5.0	626"	197°; Optical
NGC 5824	GC	15 04.0	−33 04	9.0	6'	
NGC 5822	OC	15 05.2	−54 21	6.5p	40'	
Kappa	**	15 11.9	−48 44	3.9,5.8	27"	144° (1951)
Zeta	**	15 12.3	−52 06	3.4,7.0	72"	249° (1938)
Delta	**	15 21.3	−40 39	3.2,6.2	360"	Optical
NGC 5927	GC	15 28.0	−50 40	8.3	12'	
B228	Dk	15 45.5	−34 24		240' × 20'	
NGC 5986	GC	15 46.1	−37 47	7.1	10'	
Δ 192	**	15 47.1	−35 31	7.2,7.5	35"	143° (1938)
SL 11	Dk	15 57.0	−37 48		40' × 5'	
SL 7	Dk	16 01.8	−41 52		160' × 10'	

Lupus, the Wolf, is a little-known constellation to the residents of the northern hemisphere. Its faint stars are found to the south of Libra, between Centaurus and Scorpius. With the plane of our galaxy skirting its eastern border, Lupus hosts a nice collection of open clusters, globular clusters, and double stars for binoculars.

Tau[1+2] Lupi forms a wide optical double that is easy to spot in all binoculars. Look for this unrelated pair about 3° northeast of Alpha

Lupi. Tau[1] glows at 5th magnitude, while Tau[2] shines a full magnitude brighter. Both appear perfectly white. Sharp-eyed observers with 7× and larger glasses might also detect a 9th-magnitude star between Tau[1] and Tau[2], giving the impression that Tau is a triple star system. In reality, this tiny star is gravitationally bound to Tau[1], but neither it nor Tau[1] is linked to Tau[2].

NGC 5822 is a large, attractive open cluster found along the Wolf's southern border. With an apparent diameter exceeding that of the Full Moon, this is an ideal open cluster for binoculars. Through 11 × 80 glasses, I recorded it as a "very large, moderately dense collection of many magnitude 9 and magnitude 10 stars." NGC 5822 is about 1,800 light years distant.

As you look toward NGC 5822, you will undoubtedly notice a second open cluster to the south. This is NGC 5823, described under its home constellation, Circinus.

Barnard 228 is a long, thin band of dark nebulosity that stretches for 4° diagonally from northwest to southeast across Lupus. Due to its broad expanse, Barnard 228 is more easily seen through lower power, wide-field glasses. Look for it about halfway along and running perpendicular to a line connecting Psi Lupi and Chi Lupi.

NGC 5986 is the largest of four globular clusters visible in binoculars within Lupus. Its circular disk is found to the west of Eta Lupi, nestled within a triangle of 6th- and 7th-magnitude stars. I found it plainly evident through 7× glasses.

Lynx (Lyn)

Object	Typ	R.A. (2000) h m	Dec. (2000) ° '	Mag.	Size/Sep/ Period	Comments
5	**	06 26.8	+58 25	5.3,7.9	96"	272° (1924); 5036
R	Vr	07 01.3	+55 20	7.2–>14.5	378.66 days	Long Period Variable
SV	Vr	08 03.7	+36 21	8.2–9.1p	70 days	Semi-Regular
NGC 2683	Gx	08 52.7	+33 25	9.7	8' × 1'	Sb

Lynx occupies a rather bare portion of the early spring northern sky beyond Gemini. This cat is faint, with its brightest stars shining at no more than 4th magnitude. While many deep sky objects are found within the constellation, most are galaxies too feeble to be visible through most glasses. For binocular observers who take the time to wade through the "missing" Lynx, four deep sky objects await.

5 Lyncis is a binary star found near the constellation's northern limit. As seen through most glasses, the 5th-magnitude A star is joined by an 8th-magnitude partner (not the B star; see shortly). With a separation measuring in excess of 1′ of arc, both suns shine forth clearly.

In reality, 5 Lyncis is a triple star system: 5 Lyncis B shines at magnitude 9.8 and is placed about a half arc-minute away from the primary sun. This combination makes it a difficult catch through even giant glasses.

NGC 2683 is the only galaxy in Lynx that is faintly visible through 7× glasses. To find this Sb spiral, move to the southeastern sector of Lynx, and look about 2° north-northwest of a pleasant little asterism involving Sigma[1] and Sigma[2] Cancri. Inclined nearly edge on to our line of sight, NGC 2683 presents a long, thin disk punctuated by a brighter oblong nucleus. Visually, this galaxy spans about 6′ × 1′, but large-telescope photographs increase its dimensions to about 8′ × 1′.

Lyra (Lyr)

Object	Typ	R.A. (2000) h m	Dec. (2000) ° ′	Mag.	Size/Sep/ Period	Comments
W	Vr	18 14.9	+36 40	7.3–13.0	196.54 days	Long Period Variable
T	Vr	18 32.3	+37 00	7.8–9.6		Irregular
Epsilon[1+2]	**	18 44.3	+39 40	5.0,5.2	208″	173° (1955); 11635; Double-Double
Zeta	**	18 44.8	+37 36	4.3,5.9	44″	150° (1955); 11639
Isk 1	OC	18 48	+37		110′	
Beta	**	18 50.1	+33 22	3.3v,8.6	46″	149° (1955); 11745; A = mag 3.3–4.3
Steph 1	OC	18 53.5	+36 55	3.8	20′	Delta Lyr cluster
NGC 6720	PN	18 53.6	+33 02	9.7	70″ × 150″	M57 Ring Nebula
Delta[1+2]	**	18 54.5	+36 54	5.6,4.5	630″	Optical
OΣ 525	**	18 54.9	+33 58	6.0,7.7	45″	350° (1935); 11834
R	Vr	18 55.3	+43 57	3.9–5.0	46.0 days	Semi-Regular
NGC 6779	GC	19 16.6	+30 11	8.2	7′	M56
OΣΣ 181	**	19 20.1	+26 39	7.6,7.4	58″	3° (1923)
NGC 6791	OC	19 20.7	+37 51	9.5	16′	
RR	Vr	19 25.5	+42 47	7.1–8.1	0.567 day	RR Lyrae prototype

With its origin dating back to the time of Homer, the constellation **Lyra** rides near the zenith on warm summer nights in the northern hemisphere. Lyra represents a lyre—a small, harplike instrument. Four faint stars set in a parallelogram frame the harp's body, while brilliant Vega (a type-A star and fifth brightest in the sky) marks a portion of the handle.

Although the constellation is rather small, a wide variety of objects are packed within. Some we have all heard of, while others are known to few amateurs.

Epsilon^{1+2} Lyrae, better known as the Double-Double, is one of the most popular double stars in the summer sky. Epsilon's duality is evident with only the slightest optical aid and may even be faintly glimpsed with the unaided eye just northeast of dazzling Vega. The stars are separated by 208″ of arc, which is near the naked eye resolution limit for most people. The northernmost of the pair is labeled Epsilon^1, while the other is called Epsilon^2.

The reason for the name "Double-Double" becomes self-evident in 3-inch and larger telescopes at about 100×. If the seeing is good, Epsilon^1 and Epsilon^2 should be further resolved into two tight pairs. Epsilon^1 consists of a 6th-magnitude secondary star orbiting a 5th-magnitude primary star, while Epsilon^2 is formed from a magnitude 5.3 secondary star found nearly due east of the brighter, magnitude 5.1 primary. Astronomers have classified all four as spectral type-A stars, with Epsilon^1 A also thought to be a spectroscopic binary.

Zeta Lyrae pinpoints the parallelogram's northwestern corner. This wide double star is composed of magnitude 4.5 and 5.6 suns separated by about 44″ of arc, with the fainter companion found due south of the primary. Observations made across the years indicate that both are gravitationally bound to each other and form a true binary system.

Beta Lyrae lies at the southwestern corner of the Harp's body and is truly a star for all observers. On the one hand, Beta is a challenging double star, with magnitude 3.3 and 8.9 white components severed by about 46″ of arc. On the other hand, Beta A itself is an eclipsing binary, with an unseen third star causing the light of Beta A to fluctuate from magnitude 3.3 to 4.3 across a 12.940-day period. The Beta Lyrae family of stars does not stop here either. Although too faint to be seen through binoculars, four additional stars belong to this system. Beta Lyrae is, at the very least, a *septuple* star! You will find a chart for Beta Lyrae in Chapter 6.

Stephenson 1 is a wide, rather scantily packed open cluster of about 15 stars. It would probably pass unnoticed were it not for its brightest member, 4th-magnitude **Delta Lyrae**[1+2], the northeastern star in the Lyra parallelogram. Binoculars aimed at the cluster easily reveal both blue-white Delta[1] and orangish Delta[2], shining at magnitudes 5.5 and 4.5, respectively. The two are separated by better than 10′ of arc and are surrounded by eight to ten fainter cluster members scattered across 20′ toward the west. The view is especially nice through 10× and larger glasses, whose greater light-gathering ability allows more of the fainter stars to be seen. The distance to both Delta Lyrae and Stephenson 1 is believed to be about 1,050 light years.

M57 (NGC 6720), the Ring Nebula, visible in binoculars? You bet! Of course, we cannot expect to see the Ring as we might in a telescope, since binoculars simply do not have the magnification required to resolve its classic smoke-ring shape. However, even through light-polluted suburban skies, it is visible in 7 × 35 glasses as a faint, ever-so-slightly fuzzy "star." Increasing the magnification to about 15× begins to clearly show that this is no ordinary "star," while through 25× to 30× "giant giants," the smoke-ring shape just begins to reveal itself. This is a much better view than Messier ever had of his catalogue's 57th entry. He suspected that it was composed of very faint stars, although they could not be distinguished. It was not until William Herschel observed M57 that its true nebulous nature began to come to light. Herschel reported it as "one of the curiosities of the heavens," an apt description indeed.

We now know that the Ring is a planetary nebula, the remnant of a long-ago nova. The star that produced the Ring is still visible, although only faintly, through the largest backyard telescopes. For binocularists, however, the thrill is simply in seeing M57. Look for it about two-fifths of the way from Beta Lyrae to Gamma Lyrae, along the southern side of the Lyra parallelogram.

R Lyrae is a great variable star for study through low-power binoculars. Classified as a type-M red semi-regular star, it fluctuates between magnitudes 3.9 and 5.0 with a period of about 46 days. Look for R Lyrae in the northern part of the constellation, about 6° northeast of Vega (see the chart for Beta Lyrae in Chapter 6).

M56 (NGC 6779), discovered by Messier in 1779, is set among a well-populated field of stars about halfway between Beta Cygni (Albireo) and Gamma Lyrae. Binoculars disclose it as a fuzzy 8th-magnitude point featuring a brighter core. Resolution of the cluster's stars does not begin through instruments with less than an 8-inch aperture.

RR Lyrae is another variable suitable for binoculars. With great precision, this star rises to magnitude 7.1, falls to magnitude 8.1, and then rises back up again in a 13-hour, 36-minute time period. RR Lyrae heads up a class of ancient variable stars known as cluster variables, so called because many are found within globular clusters. All RR Lyrae stars have approximately the same absolute magnitude (about +0.5). By knowing both the absolute and apparent magnitudes, the distance to the star (and therefore to its home globular cluster) is easily derived.

Mensa (Men)

Object	Typ	R.A. (2000) h m	Dec. (2000) ° ′	Mag.	Size/Sep/ Period	Comments
U	Vr	04 09.6	−81 51	8.0–10.9p	407 days	Semi-Regular
TZ	Vr	05 30.2	−84 47	6.2–6.9p	8.569 days	**Eclipsing Binary**

Mensa, the Table, lies in the far southern skies of autumn and is seen well only from points below Earth's equator. Lacaille devised this obscure, little constellation in 1752, naming it after Table Mountain near Cape Town, South Africa.

Our attention is drawn to Mensa mainly because the southern perimeter of the Large Megallanic Cloud invades its northern border. However, while the portion of the Large Magellanic Cloud found in neighboring Dorado is rich in deep sky objects, the small wedge within Mensa yields no one particular object that is outstanding in binoculars. Instead, we can sit back and marvel at the overall wonder of the Large Magellanic Cloud. This does not mean that the rest of Mensa is totally without interest to binocularists. Far south of the Large Magellanic Cloud is a pair of "local" variable stars suitable for binocular study.

TZ Mensae is the brighter and more southern of the above-mentioned two variables. Tucked just inside the far southern border of Mensa, TZ, an eclipsing binary, fluctuates between about 6th and 7th magnitude over 8½ days.

Microscopium (Mic)

Object	Typ	R.A. (2000) h m	Dec. (2000) ° '	Mag.	Size/Sep/ Period	Comments
T	Vr	20 27.9	−28 16	7.7–9.6p	344 days	Semi-Regular
Iota	**	20 48.5	−43 59	5.1,7.0	180″	Optical
Δ 236	**	21 02.2	−43 00	6.5,6.9	57″	73° (1951)
S	Vr	21 26.7	−29 51	7.8–14.3	208.93 days	Long Period Variable

With its brightest stars only of the 5th magnitude, **Microscopium,** the Microscope, appears as little more than empty space to the unaided eye. The celestial microscope's location is easy to pinpoint though, as the Capricornus "arrowhead" is aimed directly at its center. Only a few scattered deep sky objects entice the observer with binoculars to explore the region.

T Microscopii is tucked just inside Microscopium's northwestern corner. Larger glasses have little trouble revealing its presence across the entire 344-day period during which it fluctuates between photographic magnitudes 7.7 and 9.6. Look for this long period variable to the north of a crooked rectangle of half a dozen 6th-magnitude stars.

Iota Microscopii appears through most glasses as a tight asterism of three stars. Iota is formed from a 5th-magnitude primary sun teamed with a 7th-magnitude companion separated by 3′ of arc. The third star in the system shines at magnitude 6.5 and lies about 13′ to the south. All three stand out well against the stark surroundings.

Dunlop (Δ) 236 is the central and brightest of three stars located about halfway between Iota and Eta Microscopii. Even the least powerful glasses should have little trouble resolving both members of this stellar pair, as they shine at 6th and 7th magnitude and are separated by nearly a full minute of arc.

Monoceros (Mon)

Object	Typ	R.A. (2000) h m	Dec. (2000) ° '	Mag.	Size/Sep/ Period	Comments
NGC 2215	OC	06 21.0	−07 17	8.6	11'	
Cr 91	OC	06 21.7	+02 22	6.4p	17'	
V	Vr	06 22.7	−02 12	6.0–13.7	333.80 days	Long Period Variable
Cr 92	OC	06 22.9	+05 07	8.6p	11'	
T	Vr	06 25.2	+07 05	5.6–6.6	27.021 days	Cepheid
NGC 2232	OC	06 26.6	−04 45	3.9	30'	
NGC 2236	OC	06 29.7	+06 50	8.5	7'	
Cr 96	OC	06 30.3	+02 52	7.3	8'	
Cr 95	OC	06 30.5	+09 56		19'	
Cr 97	OC	06 31.3	+05 55	5.4	21'	
NGC 2237	DN	06 32.3	+05 03		80' × 60'	Rosette Nebula
NGC 2244	OC	06 32.4	+04 52	4.8	24'	Rosette Nebula cluster
NGC 2251	OC	06 34.7	+08 22	7.3	10'	
NGC 2252	OC	06 35.0	+05 23	8.0p	20'	
Cr 106	OC	06 37.1	+05 57	4.6p	45'	
Cr 107	OC	06 37.7	+04 44	5.1	35'	
Hrr 5	OC	06 41	−09		15'	Asterism
Cr 111	OC	06 38.7	+06 54	7.0p	3'	
NGC 2261	DN	06 39.2	+08 44	10.0	2'	Hubble's Variable Nebula
NGC 2264	OC	06 41.1	+09 53	3.9	20'	Christmas Tree cluster
Do 25	OC	06 45.1	+00 18	7.6	24'	
NGC 2286	OC	06 47.6	−03 10	7.5	15'	
NGC 2301	OC	06 51.8	+00 28	5.8	12'	
NGC 2302	OC	06 51.9	−07 04	8.9	3'	
X	Vr	06 57.2	−09 04	6.9–10.0	155.70 days	Semi-Regular
NGC 2316	DN	06 59.7	−07 46		4' × 3'	
NGC 2323	OC	07 03.2	−08 20	5.9	16'	M50
NGC 2324	OC	07 04.2	+01 03	8.5	8'	
NGC 2335	OC	07 06.6	−10 05	7.2	12'	
RY	Vr	07 06.9	−07 33	7.7–9.2	466 days	Semi-Regular
NGC 2343	OC	07 08.3	−10 39	6.7	7'	
NGC 2353	OC	07 14.6	−10 18	7.1	20'	
U	Vr	07 30.8	−09 47	6.1–8.1p	92.26 days	RV Tauri type
UX	Vr	07 59.3	−07 30	8.0–8.9	5.905 days	Eclipsing Binary
NGC 2506	OC	08 00.2	−10 47	7.6	7'	
Zeta	**	08 08.6	−02 59	4.3,7.8	67"	245° (1936); 6617

One of the regions most abundant in deep sky objects in the winter is located within the large triangular void framed by the brilliant stars Sirius, Betelgeuse, and Procyon. This seemingly empty portion of the seasonal sky is actually the constellation **Monoceros,** the Unicorn.

With its brightest star shining at only a poor 4th magnitude, Monoceros commands very little attention from naked eye skywatchers. The constellation's origins are traceable to 16th-century Persia, although the Unicorn was not widely recognized until German mathematician Jakob Bartsch included it on his star charts of 1624. Whatever Monoceros lacks in naked eye glory it more than makes up for in its many beautiful treasures buried within.

Collinder 91 just makes it into the Unicorn's western border with Orion. Twenty stars fall within its 17' diameter, giving the cluster an overall magnitude of 6.4. Only four of its suns are bright enough to be seen separately through binoculars. This poor concentration has led some astronomers to conclude that this diamond-shaped stellar array is actually nothing more than a chance asterism.

NGC 2232 mysteriously remained unplotted on many star atlases for years, although it is easily seen in binoculars and may even be spotted with the unaided eye. The brightest star found within this open cluster is **10 Monocerotis,** at 5th magnitude. Most of its other 20 suns range between 8th and 10th magnitude and are set in a triangular pattern. With an apparent diameter equal to that of the Full Moon, NGC 2232 is actually seen better through binoculars than through most telescopes.

Collinder 96 is another little-observed member of the Collinder family of open clusters. With an overall magnitude of 7.3, the cluster's brightest stars are no greater than magnitude 8.8. This can make spotting it through 7× binoculars a bit tentative. Through 11 × 80 glasses, it impresses me as a weak semicircular harvest of stars measuring about 8' across.

Collinder 97 proves to be a rather poorly defined group. Fifteen stars are thought to belong to this open cluster, but only four or five are bright enough to be easily seen through binoculars. The brightest star in Collinder 97 is **AX Monocerotis,** an odd type-B star that fluctuates from magnitude 6.6 to 6.9 with a 232-day period. AX is unique in that it does not seem to fit into any established category of variable stars. Clearly, more study is required.

Figure 7.16
The bright open cluster NGC 2244 surrounded by the Rosette Nebula, NGC 2237, in Monoceros. This photograph was taken by Johnny Horne using an 8-inch f/1.5 Schmidt camera, hypered Technical Pan 2415 film, and a six-minute exposure.

NGC 2244 is one of the most attractive open clusters in the winter sky. Binoculars reveal half a dozen 6th- and 7th-magnitude stars set in a distinctly rectangular pattern and highlighted by the yellowish glint of 6th-magnitude 12 Monocerotis. In all, about 40 suns are affiliated with this 2,445-light-year-distant group.

Under crystalline skies, steadily held binoculars might reveal a faint haze encircling NGC 2244. This dim, irregular glow is all that may be captured visually of the famous Rosette Nebula (Figure 7.16). Most deep sky catalogues list the Rosette as **NGC 2237**, but it was originally recorded as four separate entries: NGCs 2237, 2238, 2239, and 2246. These refer to the four brightest portions of this huge cloud. Collectively, they form a wreathlike cloud spanning more than a degree across.

My best view of the Rosette complex came on one especially clear winter night a few years ago. Using my 11×80 binoculars, I could see a soft, broken ring surrounding the cluster stars of NGC 2244. Nebula filters are recommended to see the Rosette at its best, but are not absolutely necessary if skies are extremely clear.

Collinder 106 is a widespread stellar group found about 2° northeast of the Rosette. Twenty stars populate its ¾° area and combine to give this object an overall magnitude of 4.6. About a dozen of these stars are visible through binoculars, with the three brightest forming a neat triangle along the cluster's southern edge.

Collinder 107, positioned to the Rosette's east, appears as a bright triangular pattern of six stars amidst the faint glow of dimmer cluster members. Binoculars are used to best advantage for viewing this group, as it spans over half a degree.

Harrington 5 is one of "my" deep sky objects that I would like to pass along for readers' inspection. This arrowhead-shaped asterism of half a dozen 8th-magnitude stars may be found along the southern border of Monoceros, about 3° southeast of Beta Monocerotis. Though not officially labeled an open cluster, it stands out in larger glasses nicely against the rich surroundings.

NGC 2261 is a strange and marvelous celestial specter known popularly as Hubble's Variable Nebula. This is the brightest tuft of a monstrous diffuse cloud that spans over 12 square degrees. The shape of NGC 2261 will remind observers of a comet, with a broad "tail" extending to the north from a bright "coma," where the cloud's illuminating star, **R Monocerotis**, is located. Although quite small, NGC 2261 is easily seen in 7× glasses thanks to its high intrinsic brightness.

R Monocerotis is an irregular variable star that typically shines around 10th magnitude, but has been seen to dip to nearly 14th magnitude at times. Due to the brightness of the surrounding nebulosity, R Monocerotis remains a difficult star to study and observe.

NGC 2264 lies within the same huge patch of diffuse nebulosity as does NGC 2261. Nicknamed the Christmas Tree Cluster for its distinctive stellar pattern, this large, brilliant open cluster is easily found in all binoculars. Its brightest star, 5th-magnitude 15 Monocerotis, may be seen with the unaided eye, while nearly all of its other 20 suns are visible in 10× glasses. All fall into a cone-like shape reminiscent of a seasonal evergreen.

NGC 2301 is a striking open cluster of 80 stars. Binoculars immediately reveal a north-south string of six 8th- and 9th-magnitude stars marked at midspan by a hazy triangular clump of many stars too faint to resolve. I find this peculiar shape strongly reminiscent of a bird in flight, and like to refer to this as the "Great Bird of the Galaxy" cluster. The bird's two wings are represented by the string of stars extending from the three-sided body. Anyone care to take an ink blot test?

M50 (NGC 2323) is the only member of the famous Messier catalogue located within Monoceros. Cassini was first to set eyes upon this lovely open cluster in about 1711, while Messier stumbled upon it independently 63 years later. To find M50, take aim about one-third of the way from Sirius to Procyon. Although it appears to be located in the middle of nowhere, M50 is surprisingly easy to find in 7× binoculars and may even reach naked eye visibility on dark evenings. Of the 80 stars within, only about five or six are bright enough for viewing with binoculars, while the remainder blend into a pleasant background glow. A striking red cluster star prominently set amidst an ocean of blue-white suns highlights the scene.

Musca (Mus)

Object	Typ	R.A. (2000) h m	Dec. (2000) ° '	Mag.	Size/Sep/ Period	Comments
S	Vr	12 12.8	−70 09	5.9–6.4	9.66 days	Cepheid
NGC 4372	GC	12 25.8	−72 40	7.8	19'	
NGC 4463	OC	12 30.0	−64 48	7.2	5'	
BO	Vr	12 34.9	−67 45	6.0–6.7		Irregular
Harvard 6	OC	12 37.9	−68 28	10.7p		
R	Vr	12 42.1	−69 24	5.9–6.7	7.477 days	Cepheid
NGC 4815	OC	12 58.0	−64 57	8.6	3'	
NGC 4833	GC	12 59.6	−70 53	7.4	14'	
Harvard 8	OC	13 18.8	−67 12	9.5	4'	
NGC 5189	PN	13 33.5	−65 59	10.3p	153"	

Musca, the Fly, lies in the southern circumpolar sky immediately south of Crux. When first represented on Bayer's charts of 1603, the area was shown as Apis, the Bee. By the time Lacaille drew his charts in 1763, the region had assumed its present identity as a fly. With the galactic equator passing just to its north, Musca offers some fine star fields for amateurs to scan.

NGC 4372 may be found to the southwest of Gamma Muscae; a 7th-magnitude yellowish field star lying just to the northwest of this globular cluster will aid in its location. Shining with the brightness of an 8th-magnitude star, NGC 4372 is displayed through binoculars as a nebulous disk with a brighter core. If, however, a thin haze is present, the star's yellow light may scatter enough to prevent the cluster from being detected.

NGC 4463 is easier to observe than its size might indicate, for this open cluster lies on the southwestern shore of that great black abyss, the Coalsack. Through 7× glasses, NGC 4463 is distinguishable only as a dim spot of light. Increasing to 11× will reveal the cluster's two brightest members, surrounded by the haze of other, unresolvable stars.

NGC 4815 is another open cluster that is not as difficult to spot as its small size implies. Positioned directly in front of the Coalsack, the cluster's stars blend into a tiny, cloudlike haze. So feeble is the light from these distant luminaries that individual points of light cannot be made out without larger telescopic equipment.

NGC 4833 is a large, moderately faint globular cluster found less than a degree northwest of 4th-magnitude Delta Muscae. Lacaille, who first discovered this cluster in 1752, described it as resembling a small, faint comet. This description holds true for the group's appearance through modern binoculars as well. Resolution is possible only through 4-inch and larger telescopes, although a lone 9th-magnitude star may be seen superimposed on the cluster's northern edge. The projected distance to NGC 4833 is 18,000 light years.

NGC 5189 is a moderately bright, easy-to-spot nebula in eastern Musca. Look for it about 5° northeast of Beta Muscae and just north of a 7th-magnitude field star. Upon first spotting it in 1835, John Herschel proclaimed it to be a "very strange object." This is still the case today, for the true nature of this little cloud has been the subject of much controversy over the years. Some authorities believe NGC 5189 to be a gaseous nebula, while others argue that it is a planetary nebula. After extensive analysis, the latter view appears to have prevailed. The distance to NGC 5189 is estimated to be about 2,600 light years.

Binoculars show the nebula as a slightly oval disk immersed in a striking star field. Some keen-eyed observers can distinguish a bluish tint to the nebula, but the 14th-magnitude central star is well below binoculars' threshold of visibility.

Norma (Nor)

Object	Typ	R.A. (2000) h m	Dec. (2000) ° '	Mag.	Size/Sep/ Period	Comments
NGC 5925	OC	15 27.7	−54 31	8.4p	15'	
R	Vr	15 36.0	−49 30	6.5–13.9	492.74 days	Long Period Variable
T	Vr	15 44.1	−54 59	6.2–13.6	242.56 days	Long Period Variable
Cr 292	OC	15 50.7	−57 40	7.9p	16'	
Ru 113	OC	15 57.2	−59 28		45'	
NGC 6067	OC	16 13.2	−54 13	5.6	13'	
SL 8	Dk	16 14.2	−44 04		25' × 5'	
Cr 299	OC	16 18.4	−55 07	6.9p	20'	
NGC 6087	OC	16 18.9	−57 54	5.4	12'	
S	Vr	16 18.9	−57 54	6.1–6.8	9.754 days	Cepheid
Harvard 10	OC	16 19.9	−54 59		30'	
Ru 116	OC	16 23.6	−52 00		5'	
NGC 6134	OC	16 27.7	−49 09	7.2	7'	
NGC 6152	OC	16 32.7	−52 37	8.1p	30'	
NGC 6169	OC	16 34.1	−44 03	6.6p	7'	Mu Normae cluster
NGC 6167	OC	16 34.4	−49 36	6.7	8'	

Norma, the Carpenter's Square, was devised by Lacaille in 1752 and spans the zone between Scorpius and Triangulum Australis. Although even its brightest stars are inconspicuous to the naked eye, Norma possesses a number of deep sky objects for the backyard astronomer to marvel at.

Ruprecht 113 is a large, somewhat ill-defined open cluster in southern Norma. Spanning a diameter larger than the Full Moon, this group displays some 15 magnitude 8 and 9 suns. Look for this little-observed stellar family about 1° north of a distinctive four-star asterism which, in turn, is about 3° north of Beta Trianguli Australis.

NGC 6067 lies just north of 5th-magnitude Kappa Normae. Most binoculars reveal this open cluster as a distinct, though small, circular glow with two or three faint points of light set within. Giant glasses may resolve a few more individual stars, but most of the cluster's 100 luminaries fall below the limit of detection. Set within a stunning field sprinkled with stardust, NGC 6067 proves to be one of the unsung showpieces of the southern summer sky.

NGC 6087 also falls into that class of little-known-but-still-striking open clusters. With 8 of its 40 suns shining at 9th magnitude or brighter, this group is easily isolated with its rich environs. The cluster's brightest star is catalogued as **S Normae,** a Cepheid variable whose golden radiance varies between magnitudes 6.1 and 6.8 over a ten-day cycle.

Harvard 10 is known to few amateurs, yet it is detectable in 7× binoculars. About one-third of its 30 stars are visible in most glasses, the remainder blending together to form a nebulous background glow.

Superimposed on the southwest corner of Harvard 10 is **Collinder 299,** a second open cluster. Although a couple of its stars are resolvable in binoculars, I have never actually gotten the feeling of seeing an open cluster when I look its way. Perhaps you will do better.

NGC 6152 is a coarse open cluster spanning about ½° across. Although it is comprised of some 70 individual stars, only a faint glow splashed with half a dozen stellar specks will generally be seen through binoculars. NGC 6152 is most easily found by first locating Epsilon[1] and Epsilon[2] Arae and then scanning about 4° to their west.

Octans (Oct)

Object	Typ	R.A. (2000) h m	Dec. (2000) ° ′	Mag.	Size/Sep/ Period	Comments
Gamma[1+2+3]	**	00 10	−82	5,6,5	60′	Optical
R	Vr	05 26.1	−86 23	6.4–13.2	405.57 days	Long Period Variable
X	Vr	10 26.2	−84 21	8.7–12.7p	205 days	Semi-Regular
U	Vr	13 24.5	−84 13	7.1–14.1	302.57	Long Period Variable
Pi[1+2]	**	15 02	−83 14	5.7,5.7	18′	Optical
S	Vr	18 08.7	−86 48	7.3–14.0	258.94 days	Long Period Variable
Mel 227	OC	20 12.1	−79 19	5.3p	50′	
Sigma	*	21 08.7	−88 57	5.5	—	Southern pole star

Home of the south celestial pole, the constellation **Octans,** the Octant, was first mentioned by Lacaille in 1752. While northern observers enjoy 2nd-magnitude Polaris at the north celestial pole, southern observers must strain to see their stellar complement. Presently, the Earth's south pole is most closely aimed toward Sigma Octantis, a weak magnitude 5.5 type-A star. Thus, unless skies are crystal clear, the sky's other pole appears barren to the naked eye.

Like Ursa Minor, Octans holds primarily double stars and variable stars for amateurs to view, but unlike its northern counterpart, the Octant also contains an errant open cluster.

Pi[1] and **Pi**[2] **Octantis** combine into a wide optical double star for far southern observers. Though actually unrelated in space, both stars shine at an identical magnitude 5.7. Just north of Pi[2] (the northernmost of the pair) is a small triangle of 8th-magnitude suns, adding a bit of "accent" to the scene.

Melotte 227 is the southernmost open cluster visible from Earth. Spanning 50' of arc, 40 stars have been identified as members of this congregation, but only one in five is bright enough to be seen in binoculars. The remainder fall below 10th magnitude. Look for Melotte 227 about 3° southwest of Alpha Octantis.

Ophiuchus (Oph)

Object	Typ	R.A. (2000) h m	Dec. (2000) ° '	Mag.	Size/Sep/ Period	Comments
Rho	**	16 25.6	−23 27	5,8,7	151",156"	0°, 253° (1925); 10049
V	Vr	16 26.7	−12 26	7.3–11.6	297.99 days	Long Period Variable
Chi	Vr	16 27.0	−18 27	4.2–5.0		Irregular; Gamma Cas type
NGC 6171	GC	16 32.5	−13 03	8.1	10'	M107
NGC 6218	GC	16 47.2	−01 57	6.6	15'	M12
NGC 6254	GC	16 57.1	−04 06	6.6	15'	M10
NGC 6266	GC	17 01.2	−30 07	6.6	14'	M62
NGC 6273	GC	17 02.6	−26 16	7.1	14'	M19
NGC 6284	GC	17 04.5	−24 46	9.0	6'	
NGC 6287	GC	17 05.2	−22 42	9.2	5'	
BF	Vr	17 06.1	−26 35	6.9–7.6	4.068 days	Cepheid
R	Vr	17 07.8	−16 06	7.0–13.8	302.57 days	Long Period Variable
B244	Dk	17 10.1	−28 24		30' × 20'	
NGC 6293	GC	17 10.2	−26 35	8.2	8'	
NGC 6304	GC	17 14.5	−29 28	8.4	6'	
36	**	17 15.3	−26 36	5,6,8	732",208"	~280°, 315° (1905); 10417
U	Vr	17 16.5	+01 13	5.9–6.6	1.677 days	Eclipsing Binary
NGC 6316	GC	17 16.6	−28 08	9.0	5'	
B64	Dk	17 17.2	−18 32		20'	

(Continued on next page)

Oph (continued)

Object	Typ	R.A. (2000) h m	Dec. (2000) ° '	Mag.	Size/Sep/ Period	Comments
NGC 6333	GC	17 19.2	−18 31	7.9	9'	M9
B59,65–7	Dk	17 21	−27		300' × 160'	Stem of Pipe Nebula
B72	Dk	17 23.5	−23 28		30'	Snake Nebula
NGC 6356	GC	17 23.6	−17 49	8.4	7'	
NGC 6355	GC	17 24.0	−26 21	9.6	5'	
B78	Dk	17 33	−26		200' × 140'	Bowl of Pipe Nebula
53	**	17 34.6	+09 35	5.8,8.5	41"	191° (1949); 10635
NGC 6402	GC	17 37.6	−03 15	7.6	12'	M14
IC 4665	OC	17 46.3	+05 43	4.2	41'	
Cr 350	OC	17 48.1	+01 18	6.1p	45'	
RS	Vr	17 50.2	−06 43	5.3–12.3p		Recurrent nova (1967)
S 694	**	17 52.1	+01 07	6.9,7.1	82"	237° (1923)
V533	Vr	17 53.1	−02 35	8.3–9.3p	32 days	Semi-Regular
Barnard's Star	*	17 57.9	+04 24	9.5		
Mel 186	OC	18 01	+03	3.0p	240'	
NGC 6572	PN	18 12.1	+06 51	9.0p	8"	
RY	Vr	18 16.6	+03 42	7.5–13.8	150.53 days	Long Period Variable
NGC 6633	OC	18 27.7	+06 34	4.6	27'	
X	Vr	18 38.3	+08 50	5.9–9.2	334.39 days	Long Period Variable

Standing along the stream of the summer Milky Way is the large, boxy constellation of **Ophiuchus,** the Serpent Bearer. According to mythology, Ophiuchus was the physician who accompanied Jason and the Argonauts in their quest for the golden fleece. Appropriately, he is depicted in the sky as holding a serpent. The serpent has long been associated with wisdom and healing and, even today, is seen on the symbol of the American Medical Association.

Aside from his serpent, Ophiuchus bears a veritable fortune in deep sky jewels. When we look toward this constellation, our gaze is skirting along the edge of our galaxy. Not surprisingly, the area is littered with several patches of dark nebulosity, as well as many glimmering globular clusters. Indeed, many fine examples of nearly every type of deep sky object are hidden within.

Rho Ophiuchi, lying just inside the constellation's southwest border, near Antares, is a nice triple star for larger binoculars. Here, we find a 5th-magnitude primary star set in a tight triangle with 7th- and 8th-magnitude companions. All burn with white light. Two other members of the system remain unseen in binoculars, but raise Rho Ophiuchi to a quintuple star.

Rho is also surrounded by very colorful, but extremely faint, clouds of bright and dark nebulosity. The bright clouds are collectively catalogued as **IC 4604,** while the dark nebula is listed as **Barnard 42.** Although they record vividly in photographs, I do not know of any visual sightings of either.

M107 (NGC 6171) is a frequently ignored globular cluster located south of Zeta Ophiuchi. Rated magnitude 8.1, it appears as a faint, diffuse glow with a noticeably brighter core. Although M107 is one of the sky's more loosely packed globular clusters, there is no hope of resolving any of its dim suns through binoculars. Studies indicate the cluster is about 19,000 light years away.

M10 (NGC 6254) and **M12 (NGC 6218)** tie as two of the three brightest globulars found in Ophiuchus. (The third, M62, is described later.) Both are in the same binocular field, centrally placed in the hexagonal body of Ophiuchus. Messier discovered them, along with M9, M14, and M19, during the spring of 1764. He described M10 and M12 as starless nebulae.

Although their visual magnitudes are the same, the structures of M10 and M12 are quite dissimilar. On the 12-point scale used to rate the density of globulars, M10 rates Class VII, indicating moderate stellar condensation. The view through binoculars closely matches Messier's description, with a nebulous glow highlighted by an obviously brighter core. Resolution of the cluster stars, first achieved by William Herschel, is out of the question through binoculars. By contrast, M12 is classified as Class IX, indicating a looser form. In most glasses, its likeness to M10 is made different by a less evident core.

M62 (NGC 6266) rides the Ophiuchus-Scorpius border about 6° southwest of Theta Ophiuchi. This magnificent globular cluster is easily seen through binoculars as a small, fuzzy blotch of light set in an absolutely stunning star field. Long-exposure photographs trace the diameter of M62 to about 14′ of arc, but visual observers will find it only about half that size.

While most globulars appear to hover around the galactic center, M62 seems to be actually immersed in the Milky Way. Perhaps this is a clue to the cluster's irregular shape, which is much more evident in photographs than it is visually. Studies completed in 1973 indicate 89 variable stars within M62, an unusually high quantity. Most are short-period RR Lyrae stars.

M19 (**NGC 6273**) is located less than one binocular field to the northeast of M62. While most globular clusters appear perfectly round, M19 is noticeably oblate, even when viewed through low-power binoculars. One study conducted earlier this century by Harlow Shapley concluded that perhaps twice as many stars lie along the globular's major axis as along its minor length. None of the stars in M19 are resolvable through binoculars, as the brightest are only 14th magnitude.

36 Ophiuchi is formed from a fine trio of stars. Binoculars show all three of these orange suns set in a pleasant little triangle just north of a portion of the Pipe Nebula complex. The orange hues become especially vivid when their images are thrown slightly out of focus and compared to the myriad of surrounding white stars. Although unseen through binoculars, a fourth and fifth sun are also found in the 36 Ophiuchi system. All five orbs are believed to compose a true quintuple star clan.

M9 (**NGC 6333**) is located about 3½° southeast of Theta Ophiuchi and 2½° north of Xi Ophiuchi. Coming in at 8th magnitude, this globular is seen as a rather small, bright object sporting a brighter core.

Extending about 20′ to the west of the globular's edge is the crescent-shaped dark nebula **Barnard 64**. Like all of its kind, Barnard 64 is seen well only under the darkest conditions, when its contrast against the background sky is greatest. Through 11 × 80's, I have seen it as a somewhat irregular dark rift arcing 180° from east to west.

The **Pipe Nebula** is a huge dark nebula that extends for over 7° in southern Ophiuchus. Even with the unaided eye, it stands out prominently against the surrounding star-studded region. In his famous catalogue of dark nebulae, E.E. Barnard assigned this formation five separate entries. The bowl of the pipe, designated **Barnard 78**, appears as a jagged rectangular formation spanning over nine square degrees. The pipe's stem is formed from **Barnard 59, 65, 66,** and **67** and extends for over 5° to the west from the base of the bowl. With the unaided eye, the stem looks like a nearly straight dark cloud. Many of its abnormalities come to light when viewed through binoculars if skies are especially clear.

Barnard 72 is a small S-shaped dark cloud. It has been nicknamed the **Snake Nebula** for its thin, curved form that extends from the northwest corner of the Pipe Nebula's bowl. Barnard 72 curves for about 30′ of arc, making it look more like a tiny worm than a snake through 7× glasses. Eleven-power and greater binoculars are preferred, as they increase both the cloud's size and its contrast against the adjacent star fields. For best viewing, Barnard 72 requires dark, rural skies regardless of the instrument used.

M14 (NGC 6402) is the easternmost of the Ophiuchus globulars visible in binoculars. Planted in a rather empty area of eastern Ophiuchus, it is often passed over in favor of more easily found targets. What a shame, as M14 is a bright object that stands out well. Like Messier's telescope, binoculars reveal the cluster as a nebulous smudge without stars. As with so many other globular clusters, M14 was first resolved into separate points of light by William Herschel. The brightest of M14's stars are no greater than 15th magnitude.

IC 4665 is a large, striking open cluster just northeast of Beta Ophiuchi. On dark, clear nights, and with no optical aid at all, this cluster may be seen as a hazy spot measured nearly two Full Moons across. Binoculars quickly resolve IC 4665 into a fine collection of about 30 magnitude 6 and fainter blue-white constituents. Though its stellar density is not very high, IC 4665 stands out well thanks to the sparseness of its environment. A small triangle of three suns highlights the cluster's center.

Collinder 350 is a second large open cluster found near Beta Ophiuchi. Begin your search for this oft-overlooked group by scanning southward from Beta and Gamma Ophiuchi. There, about 2° south of Gamma, lies our quarry. Unlike IC 4665, however, Collinder 350 is a poor collection of only about 20 faint stars that show little sign of forming a cluster. All that distinguishes this from a random field is a north-south arc of four 8th-magnitude stars aimed at an east-west arc of three 8th-magnitude points of light. Not exactly a striking object, but still a nice addition to your "life list" of conquered deep sky objects.

Barnard's Star, famous as the second closest star to our Solar System, is found about 3½° east of Beta Ophiuchi. Appearing as a reddish magnitude 9.5 point of light, it was discovered in 1916 by E.E. Barnard. Comparing photographs taken 12 years apart, Barnard noticed a faint star that seemingly exhibited an unusually large proper motion against the backdrop of more distant stars. It is now known that Barnard's Star, a type-M red dwarf, is moving almost due north at a rate of 10.29 arc-seconds per year, the fastest proper motion of any known star.

Melotte 186 has one of the largest apparent diameters of any open cluster in our sky survey. Ironically, it is also one of the most difficult clusters to sight and confirm. The group's center is nearly superimposed on 67 Ophiuchi, with edges that trace a full 2° radius. What makes Melotte 186 so difficult to identify is determining which stars in this vast region belong to it and which do not. When viewed collectively in *very* low-power opera glasses, the area transforms itself into an attractive sight set apart from the surroundings by its many 4th- to 8th-magnitude suns.

NGC 6633 may be found very near the Ophiuchus-Serpens border. The 5th-magnitude shimmer of this large open cluster is visible with the naked eye, given good sky conditions. Binoculars beautifully display most of the 30 magnitude 8 stars in a rich 27'-diameter swarm. These stars shine with a blue-white luster against a nebulous glow from the few cluster members left unresolved. NGC 6633 is estimated to be just over 1,000 light years away and is well isolated from the starry path of the Milky Way.

Orion (Ori)

Object	Typ	R.A. (2000) h m	Dec. (2000) ° '	Mag.	Size/Sep/ Period	Comments
NGC 1662	OC	04 48.5	+10 56	6.4	20'	
W	Vr	05 05.4	+01 11	8.6–11.1p	212 days	Semi-Regular
NGC 1788	DN	05 06.9	−03 21		8' × 5'	
23	**	05 22.8	+03 33	5.0,7.1	32"	28° (1934); 3962
Cr 65	OC	05 26	+16	3.0p	220'	
S	Vr	05 29.0	−04 42	7.5–13.5	419.20 days	Long Period Variable
B30–2	Dk	05 29.8	+12 32		80' × 55'	3° NW of Lambda Orionis
Delta	**	05 32.0	−00 18	2.2,6.3	53"	359° (1932); 4134
Σ 747	**	05 35.0	−06 00	4.8,5.7	36"	223° (1924); 4182
Cr 69	OC	05 35.1	+09 56	2.8p	65'	Lambda Orionis Cluster
KX	Vr	05 35.1	−04 44	6.9–8.1p		Irregular
NGC 1981	OC	05 35.2	−04 26	4.6	25'	
42 + 45	**	05 35.4	−04 50	4.7,5.3	6'	
Theta^{1+2}	**	05 35.4	−05 25	4.9,5.0	135"	314° (1926); 4188; (Theta^1 = Trapezium)
Theta^2	**	05 35.4	−05 25	5.2,6.5	52"	92° (1937); 4188
NGC 1976	DN	05 35.4	−05 27	2.9	66' × 60'	M42 Orion Nebula
NGC 1977	DN	05 35.5	−04 52	4.6	20' × 10'	
NGC 1982	DN	05 35.6	−05 16	6.9	20' × 15'	M43; NW part of M42

Ori (continued)

Object	Typ	R.A. (2000) h m	Dec. (2000) ° '	Mag.	Size/Sep/ Period	Comments
Cr 70	OC	05 36	−01	0.4	150'	Belt stars
NGC 2024	DN	05 40.7	−02 27		30' × 30'	
B35	Dk	05 45.5	+09 03		20' × 10'	
NGC 2068	DN	05 46.7	+00 03	8	8' × 6'	M78
NGC 2071	DN	05 47.2	+00 18		4' × 3'	NE of M78
B36	Dk	05 49.7	+07 31		120'	
NGC 2112	OC	05 53.9	+00 24	8.6	11'	
Alpha	Vr	05 55.2	+07 24	0.4–1.3	2,110 days	Semi-Regular Betelgeuse
U	Vr	05 55.8	+20 10	4.8–12.6	372.40 days	Long Period Variable
V352	Vr	06 01.8	−02 21	8.5–10.0p		Irregular
NGC 2169	OC	06 08.4	+13 57	5.9	7'	
NGC 2175	OC	06 09.8	+20 19	6.7	18'	
NGC 2194	OC	06 13.8	+12 48	8.5	10'	
75	**	06 17.1	+09 57	5.4,8.5	117"	159°; 4890

Dominating the winter night is **Orion,** the Hunter. This most impressive constellation blazes forth with an unparalleled array of dazzling stars seemingly suspended in an ocean of black ink. Within and surrounding the four bright stars that frame Orion's torso—Betelgeuse, Bellatrix, Rigel, and Saiph—are many of the finest deep sky objects found anywhere in the heavens.

Collinder 65 is a large, bright open cluster found in extreme northern Orion. While the cluster's center is located in Orion, many of its stars bleed over into adjacent Taurus. Due to its full 2° diameter, wide-angle glasses are best suited for taking in the entire cluster. Most of the stars in Collinder 65 shine between 6th and 8th magnitude and are set within a diamond-shaped pattern.

Delta Orionis (Mintaka) is a wide, bright double star that seems tailor made for binoculars. Viewing the westernmost belt star through most glasses will unveil a 2nd-magnitude primary paired with a 6th-magnitude companion. Both are type-B stellar infernos and shine perfectly white. The companion is found nearly a full minute of arc to the primary's northeast. Given the estimated distance of about 1,500 light years from Earth, the physical separation between the two stars must be nearly half a light year. Although binoculars do not show them, a third and fourth star belong to the system: Delta C is a 13th-magnitude point of light to the southwest of Delta A, while the fourth star is a spectroscopic companion to Delta B.

Collinder 69 encircles 3rd-magnitude Lambda Orionis, the northernmost star in the Hunter's triangular "head." Joining Lambda are 19 fainter stars ranging from 5th to 9th magnitude. Especially noteworthy is a chain of three suns that stretch from Lambda southward toward unrelated Psi^1 Orionis.

NGC 1981 is a coarse open cluster found just north of the top star in Orion's sword. Seven-power binoculars have little trouble identifying eight or nine stars within the 25'-of-arc boundaries of NGC 1981, while the remaining dozen or so fainter cluster members blend their light to form a soft background glow.

M42 (NGC 1976), the Great Nebula in Orion, dominates Figure 7.17 and is one of the finest objects found anywhere in the sky. To the naked eye, the nebula is detectable as a hazy patch surrounding Theta Orionis, the middle star in the sword. With binoculars, this area unfolds into a glowing cloud of great intricacy. The Orion Nebula, a massive HII emission region, glows with a distinctively turquoise cast and may remind you of a cupped hand reaching toward the many stars that glow hotly within. Words and drawings fail to do justice to this magnificent sight, nor can the finest photographs convey the same thrill as viewing M42 firsthand. Even as an observer gains experience over the years, there will always be some aspect of M42 that has not been noticed before, adding to the experience of seeing the Orion Nebula.

Theta^1 and **Theta^2 Orionis** are found embedded within the great glowing cloud M42. Even the lowest power opera glass should have little trouble resolving them, as they shine at magnitudes 4.9 and 5.0, respectively, and are separated by better than 2' of arc. Theta^1 is the famous Trapezium quadruple star. Although bright enough to be seen through binoculars, the close proximity of these stars to one another prevents their individual resolution in less than 11× glasses. Theta^2, on the other hand, is an easily-split double star, with its magnitude 5.2 and 6.5 elements severed by almost 1' of arc.

Appearing as a detached portion of the Great Nebula is the comparatively small nebulous puff known as **M43 (NGC 1982)**. Although separable from M42 in 7× glasses, many observers never distinguish the 15' × 20' patch of M43 as an individual object. In fact, while M42 was telescopically discovered in 1610, M43 was not recognized until 1731.

The M42-M43 complex is but the brightest tip of a huge network of nebulosity that litters not just the sword, but indeed the entire constellation of Orion. North of M42, at the top of the sword,

Figure 7.17
The wondrous Orion Nebula, M42 (NGC 1976), and M43 (NGC 1982)
highlight the sword of the Hunter. Also visible, above M42, are the
diffuse nebulae NGC 1973–77–79 and the open cluster NGC 1981.
Photograph by George Viscome, who used a 3-inch f/6.6 refractor, ISO
400 film, and a 12-minute exposure.

are three separately catalogued members of Dreyer's *New General Catalogue*: NGC 1973, 1975, and 1977. Of these, only **NGC 1977** is seen clearly through binoculars. Look for its warm, irregular form engulfing four stars, including the wide double, **42** and **45 Orionis**.

Collinder 70 (Figure 7.18) is an open cluster that all of us have seen, but few are aware that it is a cluster. This is because Collinder 70 surrounds and includes all three stars in Orion's Belt. In all, 100 suns spanning 3° belong to this wide group. Most are brighter than 10th magnitude and are therefore visible in 7×50's. Although its large expanse precludes detection in the narrow fields of most telescopes, Collinder 70 is a delightful stellar family through low-power, wide-angle binoculars.

M78 (NGC 2068) is found within a spray of nebulosity about 3½° east of Mintaka along the celestial equator. Discovered by Méchain in 1780, M78 is strongly reminiscent of a faint comet. Two 10th-magnitude stars located within the "head" of this false comet give the impression of twin nuclei, while the nebula's "tail" fans out toward the southeast.

Alpha Orionis, better known as Betelgeuse, is one of the most striking stars visible in the night sky. Classified a type-M supergiant, Betelgeuse is seen as a distinctly orange or red 1st-magnitude beacon marking the right shoulder of Orion. With an estimated diameter equal to about 500 million miles (greater than the orbital diameter of Mars), it is probably the largest star visible to the unaided eye.

For visual observers, Betelgeuse is also noteworthy as the brightest star in the sky that is noticeably variable. Across a semi-regular 2,110-day cycle, its output fluctuates between magnitudes 0.4 and 1.3. This change corresponds to an actual drop in light emission of about 46%.

NGC 2169 is quite easily found to the southeast of the midpoint between Nu Orionis and Xi Orionis in the Hunter's raised right arm. This is a small, bright open cluster, consisting of about 30 stars ranging from magnitude 8 to magnitude 10. Binoculars reveal the four brightest cluster members buried in a faint, misty glow. And while you are in the area, be sure to scan the richness of the surrounding Milky Way fields.

Figure 7.18
The three belt stars of Orion are the brightest members of the little-known open cluster Collinder 70. Below is Orion's sword, highlighted by M42. Photograph by Lee Coombs using a 150-mm lens at f/5.6, ISO 125 film, and a 15-minute exposure.

Pavo (Pav)

Object	Typ	R.A. (2000) h m	Dec. (2000) ° '	Mag.	Size/Sep/ Period	Comments
Lambda	Vr	18 52.2	−62 11	3.4–4.4		Irregular; Gamma Cas type
Kappa	Vr	18 56.9	−67 14	3.9–4.8	9.088 days	Cepheid; W Virginis type
NGC 6744	Gx	19 09.8	−63 51	9.0	15' × 10'	SBb+
NGC 6752	GC	19 10.9	−59 59	5.4	20'	
T	Vr	19 50.7	−71 46	7.0–14.0	243.97 days	Long Period Variable
S	Vr	19 55.2	−59 12	6.6–10.4	386.26 days	Long Period Variable

Pavo, the Peacock, is another of the rather nondescript constellations of the deep south that was added by Bayer in his *Uranometria* atlas of 1603. Lying away from the Milky Way plane, Pavo contains only six deep sky objects of interest to binocular observers. Happily, each offers something special.

Lambda Pavonis is one of the brightest irregular variables in the entire sky. Classified as a Gamma Cassiopeiae type of star, it undergoes erratic fluctuations between magnitudes 3.4 and 4.4 that may be followed easily through all glasses and even with the unaided eye.

Kappa Pavonis may also be studied with or without optical aid. Belonging to the W Virginis, or Type II, subclass of Cepheid variables, this star changes with great precision between about 4th and 5th magnitude across a nine-day period.

NGC 6744 is one of the southern hemisphere's finest galaxies. Lying a few degrees east of a pleasant little asterism containing Theta Pavonis, NGC 6744 appears through most glasses as a gentle glow highlighted by a brighter core. Most published estimates place its visual magnitude between 8.5 and 9.0, which means that it will probably remain undetected through smaller binoculars, except on crystal clear nights. Telescopic photographs reveal NGC 6744 to be a magnificent, multiple-armed barred spiral with an abundance of detail visible throughout.

NGC 6752 is one of the premium globular clusters in the sky, yet it seems doomed to anonymity because of its southern declination. For those fortunate enough to have a good view of Pavo, NGC 6752 must surely be a favorite object. In fact, given good sky conditions, keen-eyed observers have little trouble discerning it without any optical aid at all as a dim smudge of light about 2° due east of 4th-magnitude Omega Pavonis. Through binoculars, NGC 6752 appears as a bright, perhaps slightly yellowish globe measuring about 5′ across. (Photographs increase its extent by a factor of four.) Resolution of the cluster must remain with the telescope only, as none of the approximately 100,000 member stars shine at greater than about 12th magnitude. Studies indicate this swarm to be about 14,000 light years away, making it one of the closer globular clusters to Earth.

Pegasus (Peg)

Object	Typ	R.A. (2000) h m	Dec. (2000) ° ′	Mag.	Size/Sep/ Period	Comments
85	**	00 02.2	+27 05	5.8,8.6	76″	330° (1932); 17175
NGC 7078	GC	21 30.0	+12 10	6.4	12′	M15
3	**	21 37.7	+06 37	6.0,8.3	39″	349° (1934); 15147
Epsilon	**	21 44.2	+09 52	2.4,8.4	143″	320° (1913); 15268; Enif
AG	Vr	21 51.0	+12 38	6.0–9.4	830.14 days	Z And type
TW	Vr	22 04.0	+28 21	7.0–9.2	956.4 days	Semi-Regular
NGC 7217	Gx	22 07.9	+31 22	10.2	3′ × 2′	Sb
NGC 7331	Gx	22 37.1	+34 25	9.1	10′ × 4′	Sb
R	Vr	23 06.6	+10 33	6.9–13.8	378.02 days	Long Period Variable
W	Vr	23 19.8	+26 17	7.9–13.0	344.92 days	Long Period Variable
S	Vr	23 20.6	+08 55	7.1–13.8	319.22 days	Long Period Variable
Σ 3007	**	23 22.8	+20 34	6.6,8.9	88″	311° (1956); 16713

Dominating the northern hemisphere's autumn sky is the Great Square of **Pegasus.** In mythology, Pegasus was the mighty flying horse ridden by Perseus when he rescued Princess Andromeda from the clutches of Cetus the Sea Monster. In the sky, we see only the front half of Pegasus, with the horse's body marked by the four stars forming the square. Throughout Pegasus are several fine sights to behold with binoculars, both within the Milky Way and beyond.

M15 (NGC 7078) is easily found about 4° northwest of Enif (Epsilon Pegasi), the star at the tip of Pegasus' nose. The 18th-century observer Maraldi was the first to happen upon this globular cluster, in 1746, with Messier rediscovering it in 1764.

Shining at magnitude 6.4, M15 may be seen with the unaided eye on clear nights and is considered one of the showpiece objects of the autumn sky. Most binoculars reveal it as a nebulous patch absent of stars, although some stellar resolution is barely discernible in 25× to 30× glasses if the observer uses averted vision.

M15 is unique among globular clusters in that it is the only one known to contain a planetary nebula. Called Pease 1 and discovered in 1927 by astronomers at Mount Wilson Observatory, it shines weakly at photographic magnitude 13.8 and measures only 1″ across (certainly not a binocular object!). M15 is also famous as an X-ray source, which may suggest that one or more supernova remnants are buried deep within.

AG Pegasi may also be found by first centering on Enif. The variable is located a scant 3⅓° to the northeast. AG has long been a favorite autumn variable of mine. Across an erratic period, it varies from magnitude 6.0 to magnitude 9.4, which is bright enough to permit most binocular users to follow the star's progress throughout its cycle.

Studies conclude that AG Pegasi is a member of the Z Andromedae, or symbiotic family of variable stars. Symbiotic stars are close binary systems made up of relatively cool giant suns paired with very hot companions. The variations in magnitude result from pulsations of the cool star along with some material exchange between the two companions.

NGC 7331, a fine Sb spiral, is the brightest galaxy in Pegasus. Found about 4° north-northwest of Eta Pegasi and just south of a wide double star of equal magnitude, NGC 7331 appears much like a miniaturized version of the Andromeda Galaxy. Actually, it is every bit as large as M31, but lies about 66 million light years away. (M31 is comparatively close at 2.2 million light years.) Binoculars show NGC 7331 as an oval, 9th-magnitude smudge of grayish light that diffuses rapidly away from a brighter core.

Surrounding NGC 7331 are many smaller galaxies, including the famous Stephan's Quintet. All are much too faint for binoculars.

Perseus (Per)

Object	Typ	R.A. (2000) h m	Dec. (2000) ° ′	Mag.	Size/Sep/ Period	Comments
IZ	Vr	01 32.1	+54 01	7.8–9.0p	3.688 days	Eclipsing Binary
NGC 744	OC	01 58.4	+55 29	7.9	11′	
KK	Vr	02 10.3	+56 34	6.6–7.8	Irregular	

Per (continued)

Object	Typ	R.A. (2000) h m	Dec. (2000) ° '	Mag.	Size/Sep/ Period	Comments
NGC 869	OC	02 19.0	+57 09	4.3	30'	Double Cluster (h Per)
NGC 884	OC	02 22.4	+57 07	4.4	30'	Double Cluster (Chi Per)
S	Vr	02 22.9	+58 35	7.9–11.5		Semi-Regular
NGC 957	OC	02 33.6	+57 32	7.6	11'	
Tr 2	OC	02 37.3	+55 59	5.9	20'	
NGC 1023	Gx	02 40.4	+39 04	9.5	8' × 3'	E7p; lens-shaped galaxy
NGC 1039	OC	02 42.0	+42 47	5.5	35'	M34
Rho	Vr	03 05.2	+38 50	3.3–4.0	50 days	Semi-Regular
Beta	Vr	03 08.2	+40 57	2.1–3.4	2.867 days	Eclipsing Binary; Algol
NGC 1245	OC	03 14.7	+47 15	8.4	10'	
Mel 20	OC	03 22	+49	1.2	185'	Alpha Persei Cluster
NGC 1342	OC	03 31.6	+37 20	6.7	15'	
IC 348	OC/ DN	03 44.5	+32 17	7.3	10'	Around Omicron Per
NGC 1444	OC	03 49.4	+52 40	6.6	4'	
NGC 1499	DN	04 00.7	+36 37		145' × 40'	California Nebula
NGC 1491	DN	04 03.4	+51 19		3'	
NGC 1513	OC	04 10.0	+49 31	8.4	9'	
NGC 1528	OC	04 15.4	+51 14	6.4	25'	
NGC 1545	OC	04 20.9	+50 15	6.2	18'	
NGC 1579	DN	04 30.2	+35 16		12'	
NGC 1582	OC	04 32.0	+43 51	7p	37'	
57	**	04 33.4	+43 04	6.1,6.8	116"	198° (1913)

Perseus is another member of autumn's "royal family of the north," which also includes Andromeda, Cassiopeia, and Cepheus. According to the popular legend, Perseus rescued Andromeda just as she was about to fall victim to Cetus, the sea monster. For his valor, Perseus was placed among the stars for eternity.

Sky watchers in the northern hemisphere see Perseus standing high in the autumn sky. With the Milky Way's plane passing through, Perseus is a hero to amateur observers as well, with many fine clusters and nebulae lying within the constellation's borders.

NGC 869 and **NGC 884** (Figure 7.19) are better known to most of us as the "Double Cluster," a striking pair of open clusters. Both may be observed with the unaided eye as a faint, elongated smudge of light situated about halfway between the "W" of Cassiopeia and the "tip"

Figure 7.19
The magnificent Double Cluster (NGC 869 and NGC 884) in Perseus.
George Viscome used a 500-mm lens for this photograph.

of Perseus. In fact, their combined brightness attracted the eyes of stargazers as long ago as 150 B.C., when Hipparchus first noted them. The clusters were later assigned the alter egos of h Persei and Chi Persei by early celestial cartographers.

Through binoculars, both clusters erupt into vast collections of stars set against a strikingly beautiful field. NGC 869 (the western-most cluster) consists of about 200 suns, while NGC 884 holds about 150 stars. Both are composed primarily of hot type-A and type-B supergiant, superluminous suns, with many colorful red supergiant stars also strewn across the region.

Studies indicate that the two clusters are not physically linked to one another, nor do they appear to have had a common origin. NGC 869 is estimated to lie about 7,100 light years away and is thought to be about 5.6 million years old. NGC 884 is a bit farther away from us at 7,500 light years and a bit younger, with an estimated age of 3.2 million years. Each is about 70 light years across.

Trumpler 2 is a pleasant open cluster found about 2° west of Eta Persei at the constellation's northern "tip." Twenty blue and white stars of 7th magnitude and fainter populate the group, with about half bright enough to be seen through 7× glasses. Most of the stars are gathered near the cluster's southern perimeter and set in a rectangular pattern.

M34 (NGC 1039) is an outstanding open cluster easily spotted along Perseus' western border. Credited with its discovery in August 1764, Messier described it as "a cluster of small stars a little below the parallel of Gamma Andromedae."

M34 stretches across better than half a degree and holds 60 stars within its gravitational grip. Most shine between 7th and 13th magnitude and twinkle with pure white glows. A tight central knot of stars highlights the group, while other stellar members disperse outward toward its edges. All appear bathed in the misty glow of fainter, unresolved stars. M34 lies about 1,400 light years from Earth and is indeed a grand target for binoculars.

Beta Persei, better known as Algol, is the premier eclipsing binary star of the northern sky. With precise regularity, it fluctuates between magnitudes 2.1 and 3.4 with a 2.867-day period. During its cycle, this binary's type-B primary star is alternately covered and exposed by a fainter type-G companion, causing the overall dimming and brightening effect that we witness here on Earth. The fading lasts for about 10 hours as the companion passes behind the primary along our line of sight. As the companion moves off the primary, Algol returns to peak brightness. A second, minor dip in brightness occurs as the companion passes in front of the primary.

The apparent brightness of Algol keeps it well within the range of observation of both the naked eye and binoculars across its entire cycle, thereby making it an ideal star for fledgling variable star observers to follow. To track Algol, compare it periodically with stars of known brightness in its immediate surroundings.

Melotte 20 (Figure 7.20) is not an open cluster in the truest sense of the term, but rather a loosely-bound stellar association. Also known as the Perseus OB–3 or the Alpha Persei Association, this group covers a 3° zone of sky and contains about 70 stars down to 10th magnitude, including 3rd-magnitude Mirfak (Alpha Persei). Estimates indicate that these stars are mere stellar infants, with an average age of about 51 million years. All lie 550 light years away.

Through binoculars, this magnificent region literally explodes with beauty. Many stunning blue-white starry jewels are scattered throughout the region in numerous small clumps and asterisms. Against the backdrop of the rich star clouds of the Milky Way, Melotte 20 is truly the stellar diamond mine of the northern autumn sky.

Figure 7.20
Melotte 20 embraces a fine array of stars, including Alpha Persei to the right of center. Photograph by George Viscome using a 135-mm lens, ISO 400 film, and a 15-minute exposure.

NGC 1499, the California Nebula, seen in Figure 7.21, is a paradoxical object. At times, it can be glimpsed with the unaided eye, while at others it is invisible through large amateur telescopes. The difficulty in finding the California Nebula stems from its large surface area, which spans 145′ × 40′. While its *combined* light may equal 6th magnitude, the corresponding surface brightness drops to nearly 14th magnitude due to the cloud's vast extent. Still, given ideal conditions, several keen-eyed observers have noted NGC 1499 using only a nebula filter.

Although I have never seen it without some optical aid, I have clearly noted the entire "state" in 7×50 binoculars equipped with nebula filters. Increasing to 11×80 binoculars (still with the filters in

Figure 7.21
The elusive California Nebula (NGC 1499) in Perseus. Photograph by
Martin Germano using a 135-mm lens at f/3.25 and a 25A filter, hypered
Technical Pan 2415 film, and a 50-minute exposure.

place), some of the cloud's delicate texture could also be discerned.
Without the filters, it should be noted, NGC 1499 vanished.

NGC 1528 and **NGC 1545** form a pair of open clusters in northeastern Perseus. The former consists of about 40 stars, although only one
or two are bright enough to be resolved through most common
binoculars. The remainder form a hazy glow measuring 25′ of arc
against the background sky.

NGC 1545 is somewhat smaller than NGC 1528, although I feel
it is a bit more impressive through binoculars. A collection of 20 stars
is held within an 18′ region, with about half a dozen stars brighter
than 10th magnitude. Once again, the unresolved cluster stars blend
into a faint, distant glow.

Phoenix (Phe)

Object	Typ	R.A. (2000) h m	Dec. (2000) ° '	Mag.	Size/Sep/ Period	Comments
Zeta	Vr	01 08.4	−55 15	3.9–4.4	1.670 days	Eclipsing Binary
W	Vr	01 19.9	−55 55	8.1–14.4	331.20 days	Long Period Variable
R	Vr	23 56.5	−49 17	7.5–14.4	267.86 days	Long Period Variable
S	Vr	23 59.1	−56 35	8.6–10.6p	141 days	Semi-Regular

Phoenix, devised by Bayer in 1603 and named for the mythological bird, occupies a barren portion of the southern autumn sky to the north and west of the bright star Achernar in Eridanus. Although the region is peppered with galaxies, all are very faint and lie below the limits of binoculars. Instead, we find only a handful of closer-to-home objects to view.

Zeta Phoenicis, an eclipsing binary, is the brightest of the four variable stars visible with binoculars within this constellation. Zeta varies rapidly between magnitude 3.9 and 4.4 as the speedy, unseen secondary races around the primary to complete an orbit in just 40 hours. Its quick cycle permits observers to see the star's brightness change over the course of a single night. Newcomers to variable star observation should note that, because of the subtle fluctuation in magnitude, it can be difficult to measure Zeta's change in brightness accurately. Slow, methodical comparison with neighboring stars of known, fixed brightness is the key to success. Look for Zeta about 5° northwest of Achernar in Eridanus.

Pictor (Pic)

Object	Typ	R.A. (2000) h m	Dec. (2000) ° '	Mag.	Size/Sep/ Period	Comments
R	Vr	04 46.2	−49 15	6.7–10.0	164.2 days	Semi-Regular
S	Vr	05 11.0	−48 30	6.5–14.0	426.61 days	Long Period Variable
Theta	**	05 24.8	−52 19	6.8,6.8	38"	287° (1938)
h 3822	**	05 57.2	−53 26	6.5,7.5	56"	304° (1938)
Δ 27	**	06 16.3	−59 13	6.4,8.0	40"	229° (1950)

Lacaille carved **Pictor,** the Artist's Easel, from a region of stars sandwiched between brilliant Canopus and the constellation Dorado. The Easel is never well seen by northern hemisphere residents due to both its far southern location and the faintness of its stars. Pictor's brightest sun, Alpha Pictoris, is situated in the constellation's extreme southern tip and shines at a conservative magnitude 3.7. It is visible only from latitudes below +20°. Pictor offers only a sprinkling of double and variable stars suitable as targets for binoculars.

R Pictoris resides about 3° west of a weak asterism of about 10 stars surrounding and including Eta[1] and Eta[2] Pictoris. R is known to be a red semi-regular variable with a period that averages approximately 23½ weeks. During that time, it changes from magnitude 6.7 to 10.0. This means that observers with 10× and larger glasses should be able to follow it through its entire cycle, while it will probably become lost toward minimum light through smaller binoculars.

Theta Pictoris lies in a fairly rich star field in central Pictor, about 15° directly west of brilliant Canopus. Through 7× and larger glasses, two of the three stars in this system are visible as identical 7th-magnitude points of white light. (The third star is also 7th magnitude, but lies less than 0.5″ from Theta A.) Lower power binoculars will find clean resolution quite challenging because of the stars' closeness. It is always an interesting test of an observer's talent, sky steadiness, and instrument quality to see just "how low you can go" and still be able to split a double star like Theta Pictoris. Why not try your luck on the next clear evening?

Pisces (Psc)

Object	Typ	R.A. (2000) h m	Dec. (2000) ° '	Mag.	Size/Sep/ Period	Comments
Psi[1]	**	01 05.6	+21 28	5.6,5.8	30″	159° (1832); 899
77	**	01 05.8	+04 55	6.8,7.6	33″	83° (1833); 903
RT	Vr	01 13.8	+27 08	8.2–10.4p	70 days	Semi-Regular
Z	Vr	01 16.1	+25 46	8.8–10.1p	144 days	Semi-Regular
R	Vr	01 30.6	+02 53	7.1–14.8	344.04 days	Long Period Variable

Swimming among the faint stars of the barren autumn equatorial sky is the zodiacal star group **Pisces,** the Fishes. This constellation, which contains no stars brighter than 4th magnitude, copies a large, V-shaped wedge between Pegasus and Cetus. Although it is not obvious to the naked eye, Pisces is among the oldest constellations in the sky, with its origins tracing back to ancient Babylon.

Pisces offers only a handful of double and variable stars suitable for binoculars. A quick glance at a detailed star atlas will show that the constellation is teeming with galaxies, but sadly, all are too faint to be resolved in binoculars.

Psi¹ Piscium is a challenging double star consisting of approximately 6th-magnitude suns lying east of the northern fish's figure. The stars in this true binary system are separated by only 30″ of arc, making them resolvable, although with difficulty, through tripod-mounted 7× binoculars. (Without a steady support, resolution is all but impossible.) Glasses of 11× or greater will more readily show both type-A suns, with Psi¹B located to the southeast of Psi¹A.

Piscis Austrinus (PsA)

Object	Typ	R.A. (2000) h m	Dec. (2000) ° '	Mag.	Size/Sep/ Period	Comments
Beta	**	22 31.5	−32 21	4.4,7.9	30″	172° (1952)
Δ 241	**	22 36.6	−31 40	5.8,7.6	90″	31° (1918)
H VI 119	**	22 39.7	−28 20	6.3,7.3	87″	160° (1951); 16149
Alpha	**	22 57.6	−29 37	1.2,6.5	7200″	Fomalhaut

Piscis Austrinus swims through the southern skies of autumn. Like its northern cousin Pisces, the Southern Fish consists mostly of faint stars that are difficult to detect with the naked eye. While the fish's form is somewhat ambiguous, its presence is easily detected by the constellation's brightest star, Fomalhaut.

Fomalhaut (Alpha Piscis Austrini), listed as the 18th brightest star in the sky at magnitude 1.2, is made even more conspicuous by the total absence of any nearby bright stars. Although the spectrum of Fomalhaut reveals it to be a white type-A star, many observers at mid-northern latitudes comment on its apparent reddish tinge, undoubtedly an atmospheric effect.

Although Fomalhaut is usually not listed as a true binary star, studies have identified a magnitude 6.5 red dwarf (catalogued as GC 31978) lying about 2° to the south that seems to share Fomalhaut's proper motion through space. With Fomalhaut considered to be about 23 light years away, the two stars must be separated by close to one light year! Certainly neither sun could exert any appreciable

⌐ influence on the other at that distance, but it has been
that at some time in the past, both were members of a
ρen cluster that has since dissipated.

Object	Typ	R.A. (2000) h m	Dec. (2000) ° '	Mag.	Size/Sep/ Period	Comments
NGC 2298	GC	06 49.0	−36 00	9.4	7'	
L²	Vr	07 13.5	−44 39	2.6–6.2	140.42 days	Semi-Regular
Cr 135	OC	07 17.0	−36 50	2.1	50'	
Mel 66	OC	07 26.3	−47 44	7.0	10'	
Tr 7	OC	07 27.3	−24 02	7.9	5'	
NGC 2396	OC	07 28.1	−11 44	7.4p	10'	
Bochum 5	OC	07 30.9	−17 04	7.0		
Bochum 4	OC	07 31.0	−16 57	7.3		
Z	Vr	07 32.6	−20 40	7.2–14.6	499.67 days	Long Period Variable
NGC 2414	OC	07 33.3	−15 27	7.9	4'	
NGC 2421	OC	07 36.3	−20 37	8.3	10'	
NGC 2422	OC	07 36.6	−14 30	4.5	30'	M47
NGC 2423	OC	07 37.1	−13 52	6.9	20'	
Mel 71	OC	07 37.5	−12 04	7.1	9'	
Bochum 15	OC	07 40.1	−33 33	6.3		
Haffner 13	OC	07 40.5	−30 07		15'	
NGC 2439	OC	07 40.8	−31 39	6.9	10'	
NGC 2437	OC	07 41.8	−14 49	6.1	27'	M46
NGC 2447	OC	07 44.6	−23 52	6.2	22'	M93
NGC 2451	OC	07 45.4	−37 58	2.8	45'	
NGC 2453	OC	07 47.8	−27 14	8.3	5'	
NGC 2477	OC	07 52.3	−38 33	5.7	27'	
NGC 2467	OC/ DN	07 52.6	−26 23	7.1p	16'	
NGC 2482	OC	07 54.9	−24 18	7.3	12'	
Tr 9	OC	07 55.3	−25 56	8.7	6'	
NGC 2483	OC	07 55.9	−27 56	7.6	10'	
NGC 2489	OC	07 56.2	−30 04	7.9	8'	
AP	Vr	07 57.8	−40 07	7.1–7.8	5.084 days	Cepheid
Ru 44	OC	07 59.0	−28 35	7.2	5'	
Cr 173	OC	08 04	−46	0.6p	370'	
NGC 2527	OC	08 05.3	−28 10	6.5	22'	
NGC 2533	OC	08 07.0	−29 54	7.6	4'	
NGC 2539	OC	08 10.7	−12 50	6.5	21'	
Ru 55	OC	08 12.3	−32 36	7.8	17'	
NGC 2546	OC	08 12.4	−37 38	6.3	41'	

(Continued on next page)

Pup (continued)

Object	Typ	R.A. (2000) h m	Dec. (2000) ° '	Mag.	Size/Sep/ Period	Comments
Ru 56	OC	08 12.6	−40 28		**42'**	
RS	Vr	08 13.1	−34 35	6.5–7.6	41.388 days	Cepheid
XZ	Vr	08 13.5	−23 57	8.0–10.9p	2.192 days	Eclipsing Binary
NGC 2567	OC	08 18.6	−30 38	7.4	10'	
NGC 2571	OC	08 18.9	**−29 44**	7.0	13'	
NGC 2579	OC	08 21.1	−36 11	7.5	10'	
Cr 185	OC	08 22.5	−36 10	7.8	9'	

Many readers may be familiar with the tale of the mythological ship Argo and how it was built by Argos for Jason. Jason then set sail along with a crew of 50 Argonauts in search of the golden fleece. When their voyage was over, Athena placed the ship in the sky.

Although it never rose much above the southern horizon from mid-northern latitudes, the constellation Argo Navis contained over 800 naked eye stars within its huge borders. Due to its vast extent, the constellation was divided into four component constellations by astronomers of the late 19th century. Today, these smaller constellations are known as Puppis, the Stern; Pyxis, the Compass; Vela, the Sails; and Carina, the Keel.

For dwellers of the northern hemisphere, **Puppis** is the most familiar part of the fragmented archaic constellation Argo Navis. It lies east of Canis Major, in an area notably absent of any bright stars. Although it lacks naked eye luster, Puppis is rich in deep sky treasures.

L^2 **Puppis** is a bright variable star found about 12° northeast of brilliant Canopus. Classified as semi-regular, L^2 is a vivid red type-M sun that flickers between magnitudes 2.6 and 6.2. Its remarkable brightness across this entire range makes L^2 an ideal star for study through binoculars, or even with the naked eye. Observers should note that the listed period of 140 days is merely an average; the actual time span between maxima or minima can vary greatly.

Collinder 135 lies just south of the Canis Major border. A coarse open cluster, it spans nearly a full degree and contains many bright stars, including 3rd-magnitude Pi Puppis. Binoculars also reveal the double star v^1 and v^2 Puppis, an easily resolved 4th-magnitude pair of luminaries. A single 5th-magnitude star combines with Pi and the v^1-v^2 duo to give the cluster a distinctive triangular shape. Scattered throughout are at least a dozen fainter stars that contribute to the overall charm of this little-observed group.

M47 (NGC 2422) is one of the brightest and most appealing clusters in Puppis. Its 30 stellar members range in brightness from 6th to 12th magnitude, with more than half visible in 7× glasses. All shine with the sparkle of blue-white stellar sapphires against a velvet-black background. Surrounding M47 is a grand star-filled field highlighted by the striking orange star KQ Puppis to the cluster's west.

NGC 2423 lies just to the north of M47. Seven-power glasses resolve a lone 9th-magnitude point against the blur from the light of 40 fainter stars. Overall, NGC 2423 shines at 7th magnitude and extends across 20′ of arc. Estimates place it about 2,800 light years away, nearly twice the distance to M47.

Haffner 13 appears as a small knot of stars in western Puppis. Fifteen suns are believed to belong to this group, but only seven are bright enough to be seen clearly through binoculars. Unfortunately, the star fields that enclose Haffner 13 are almost as rich as the cluster itself, making positive identification difficult. As an aid, look about 1° southwest of 3 Puppis for a close-set pair of 8th-magnitude stars, with a lone 7th-magnitude point of light just to their southwest. All three of these stars are located within the cluster.

NGC 2439 is located about 1½° due south of Haffner 13. Studies show that 80 stars join to form this 10-arc-minute-diameter open cluster, although most of the cluster's constituents are too faint to be seen in binoculars. Through most glasses, NGC 2439 appears as a 7th-magnitude glow peppered with a single bright star and a few fainter points. The brightest star in the cluster is golden **R Puppis.** Though known to be a semi-regular variable, it fluctuates with great subtlety between magnitudes 6.56 and 6.87.

M46 (NGC 2437) lies only 1° east of M47. Both open clusters were discovered by Messier in February 1771. M46 is a rich congregation of about 100 stars ranging from magnitude 9 to magnitude 13. Smaller binoculars will show only a hazy glow, while giant glasses are capable of achieving partial resolution.

Invisible in binoculars, a tiny planetary nebula lies on the cluster's northern edge. This is NGC 2438, glowing softly at 11th magnitude and measuring 65″ of arc across. Once believed to be physically associated, the cluster and nebula have since been shown to be separate entities. M46 is thought to be 5,400 light years away, while NGC 2438 is about 3,000 light years distant.

M93 (NGC 2447) lies about 1½° northwest of 3rd-magnitude Xi Puppis and directly along the galactic equator. Messier was first to chance upon this fine open cluster back in March 1781, when he described it as a "cluster of small stars . . . between Canis Major and the prow of (Argo) Navis."

M93 is a pretty sight in binoculars. Of the 80 stars that comprise this 3,586-light-year-distant flock, only about one in ten are bright enough to be evident in 7×50's. The rest mix into a dim, triangular glow.

NGC 2451 is a striking, but often ignored, open cluster in the southern part of the constellation. Of its 40 member stars, 30 shine within range of binoculars, with the remainder creating a vague background shimmer. Highlighting the scene is c Puppis, the cluster's brightest star. This red beacon stands out well amid a bevy of blue-white stellar jewels. With an apparent diameter of 45 arc-minutes, NGC 2451 is too broad to cram into a telescope's field, but it is ideal for the wider field of binoculars.

NGC 2477 is found about 1° southeast of neighboring NGC 2451 and just north of 4th-magnitude b Puppis. Pale in comparison to NGC 2451, it is still worthy of mention. With 7×, it appears as a starless smudge of grayish light about the size of the Full Moon. Increasing to 11× or higher may just begin to reveal the brightest of the cluster's 160 stars. None of the stars shine at brighter than 10th magnitude, and all require exceptionally clear nights for easy detection.

NGC 2527 is located about 4° south of Rho Puppis. Forty suns, most of 9th magnitude or below and scattered across 22' of arc, populate the open cluster. Look for a soft glow surrounding eight or nine of the brighter cluster components, including an 8th-magnitude star lying on the group's eastern edge.

NGC 2539 is another open cluster that is easy to glimpse through nearly all binoculars. The light of its 50 suns combines to magnitude 6.5, which makes it visible even without optical aid. Seven-power glasses show it as an ill-defined blotch of light apparently touching the unrelated 5th-magnitude star 19 Puppis. By switching to higher power, some of the cluster's true magnitude 9 to 11 stars suddenly appear throughout.

NGC 2546 is a loose grouping of about 40 stars ranging from magnitude 6 to magnitude 11 and strewn across 41' of arc. In larger binoculars, look for a collection of about 20 9th-magnitude stars set within a soft glow and framed by an isosceles triangle of three 7th- and 8th-magnitude suns. That "glow" is NGC 2546. As is true with nearly all large-diameter open clusters, NGC 2546 is better seen through giant binoculars than with narrow-field telescopes of "normal" focal length.

Ruprecht 56 hugs the southern border of Puppis. Earth-based observers see this open cluster as a relatively loose gathering of 40 stars of 8th magnitude and fainter spread across nearly ¾°. Although the dimmest cluster stars are undetectable in binoculars, 15 suns shine forth in 7× glasses if the cluster is high enough above the horizon. Helping to pinpoint the exact location of Ruprecht 56 is 3rd-magnitude h^2 Puppis, superimposed on the cluster's eastern border. Apparently the association is not physical, as the star seems to be only a chance foreground object.

NGC 2571 is a collection of about two dozen magnitude 9 suns lying near the eastern edge of Puppis. The combined light of this open cluster's stars equals 7th magnitude, but resolution of individual stars seems reserved for 10× and larger instruments. Look for NGC 2571 within a triangle of 7th-magnitude stars.

Pyxis (Pyx)

Object	Typ	R.A. (2000) h m	Dec. (2000) ° '	Mag.	Size/Sep/ Period	Comments
NGC 2627	OC	08 37.3	−29 57	8.4	11'	
T	Vr	09 04.7	−32 23	6.4–14.0	7,000 days?	Recurrent nova
NGC 2818	OC	09 16.0	−36 37	8.2	9'	

Pyxis is the smallest of the four constellations formed from the archaic constellation Argo Navis. Commemorating that ship's navigational compass, Pyxis lies in the southern sky of early spring, just east of our galaxy's plane. Its brightest stars, of the 3rd and 4th magnitude, can easily be overlooked by naked eye skywatchers. Still, the region is a pleasant one to scan with wide-angle binoculars, although few objects of interest are to be found.

T Pyxidis drew the astronomical world's attention to tiny Pyxis in May 1902. That month, from seemingly out of nowhere, this 14th-magnitude sun suddenly brightened to a peak magnitude of 7.3. Over the next seven months, it dropped back to its original luminosity. Further investigation found that the star had experienced a similar novalike flare 12 years earlier in 1890, but the outburst had gone unnoticed at the time.

Another flare of T Pyxidis occurred in 1920, when the star reached magnitude 6.3. This event led astronomers to christen T Pyxidis the first recurrent nova, a rare and new breed of variable star. Since then, T Pyxidis has burst forth twice more, in 1944 and again in 1966. A 19-year cycle has been suggested, which, if true, means that we are now overdue for the next flare. Observers would be wise to make it a point to check this most unusual star regularly for any activity.

Reticulum (Ret)

Object	Typ	R.A. (2000) h m	Dec. (2000) ° '	Mag.	Size/Sep/ Period	Comments
Zeta	**	03 18.2	−62 30	5.2,5.5	310"	216° (1952)
Δ 14	**	03 38.2	−59 47	7.1,8.9	58"	271° (1916)
R	Vr	04 33.5	−63 02	6.5–14.0	278.28 days	Long Period Variable
h 3670	**	04 33.6	−62 49	5.9,9.2	32"	99° (1917); test for giant binoculars

Wedged between Hydrus, Horologium, and Dorado in the far southern winter sky is the small constellation **Reticulum,** the Net. With its principal stars set in a diamond pattern, this constellation lies to the northwest of the Large Magellanic Cloud and stands out quite well in dark skies. In fact, a portion of the constellation (from Alpha to Delta) offers an interesting field of isolated bright stars when viewed through low-power binoculars.

Zeta Reticuli is a wide double star of great beauty set near the constellation's western border. Both stars have a common proper motion and are spectral type-G, the same as our own Sun. Their 5th-magnitude golden glints, separated by about 5' of arc, are quite evident through all binoculars. They shine in stark contrast against the surrounding black abyss.

R Reticuli is a fine example of a classic long period variable star for study through binoculars. Toward maximum brightness, which occurs every 278 days, R shines at close to magnitude 6.5 with a strong reddish tint. Surrounding it is a good selection of stars of fixed brightness for comparison and estimation of R's magnitude. R fades completely from view through binoculars as it heads for the 14th-magnitude minimum, but returns in about four months on its way toward the next peak.

Sagitta (Sge)

Object	Typ	R.A. (2000) h m	Dec. (2000) ° '	Mag.	Size/Sep/ Period	Comments
U	Vr	19 18.8	+19 37	6.6–9.2	3.381 days	Eclipsing Binary
Epsilon	**	19 37.3	+16 28	5.7,8.0	89″	81° (1949); 12693
Harvard 20	OC	19 53.1	+18 20	7.7	7′	
NGC 6838	GC	19 53.8	+18 47	8.3	7′	M71
S	Vr	19 56.0	+16 38	5.3–6.0	8.382 days	Cepheid
WZ	Vr	20 07.6	+17 42	7.0–15.5p	11,900 days?	Recurrent nova (1978)
Theta	**	20 09.9	+20 55	6.5,7.4	84″	223° (1949); 13442; optical

Sagitta, the Arrow, is one of the smallest constellations in the summer sky. Composed of a tiny triangle of three stars aimed toward a fourth, the celestial Arrow has been associated with several mythological arrows of ancient times. One of the most popular myths describes Sagitta as the weapon used by Apollo to slay the Cyclops. Other stories hold it to be an errant arrow shot by Sagittarius, the Archer.

The entire constellation crosses a fine star field for binocularists to just sit back and slowly scan. Throughout are many pleasant little asterisms to be seen by the inventive eye, as well as the seven deep sky objects listed here.

U Sagittae lies to the west of the Arrow, near Collinder 399 in neighboring Vulpecula. U is a fine eclipsing binary for study with binoculars. Over the course of about 3⅓ days, it changes from magnitude 6.6 to 9.2 and back again as the blue-white primary star is partially covered and uncovered by a smaller, unresolvable yellow companion.

Harvard 20 is found to the south of the four brightest stars in the Arrow. It consists of about 20 dim suns, making positive identification difficult due to the stellar richness of the surroundings. Most glasses simply reveal only two 9th-magnitude cluster members engulfed in a soft smudge of light.

M71 (NGC 6838) is an unusual object—a rich, greatly compressed cluster consisting of 12th-magnitude and fainter stars. Since its discovery in 1775, M71 has been a difficult object to classify. Many older references still designate it as a very dense open cluster similar to M11 in Scutum, but most modern authorities hold that it is a globular cluster. Regardless of its true nature, binoculars disclose only a subtle glow set within a glittering star field.

WZ Sagittae is of the rare breed of stars known as recurrent novae. Usually seen only dimly if at all, recurrent novae suddenly flare up as many as ten magnitudes in a matter of hours, remain at peak brightness for a short period, and then subside to their original, preflare brightness.

Such is the case with WZ Sagittae. In 1913, and again in 1946 and 1978, this strange star rose from about 16th magnitude to about 7th magnitude. In each instance, it took the star less than a day to attain maximum brightness and between 40 and 60 days to settle back to 16th magnitude. Astronomers have concluded from spectroscopic studies that WZ is a white dwarf star that normally shines with only ½ of the Sun's intrinsic luminosity. At peak, however, it blazes nearly 30 times brighter than our star. A period of 11,900 days has been suggested, which means that WZ Sagittae may flare again around the year 2010.

Sagittarius (Sgr)

Object	Typ	R.A. (2000) h m	Dec. (2000) ° '	Mag.	Size/Sep/ Period	Comments
B84	Dk	17 46.5	−20 11		30′ × 15′	1.6° NE 58 Oph
X	Vr	17 47.6	−27 50	4.2–4.8	7.012 days	Cepheid
NGC 6494	OC	17 56.8	−19 01	5.5	27′	M23
NGC 6514	DN	18 02.6	−23 02	8.5	29′ × 27′	M20 Trifid Nebula
B86	Dk	18 02.7	−27 50		4′	
NGC 6520	OC	18 03.4	−27 54	7.6p	6′	
NGC 6522	GC	18 03.6	−30 02	8.6	5′	
B88–9,296	Dk	18 03.8	−24 23			Regions in M8
NGC 6523	DN	18 03.8	−24 23	5.8	90′ × 40′	M8 Lagoon Nebula
B87	Dk	18 04.3	−32 30		12′	Parrot's Head Nebula
NGC 6531	OC	18 04.6	−22 30	5.9	13′	M21
NGC 6530	OC	18 04.8	−24 20	4.6	15′	M8 cluster
NGC 6528	GC	18 04.8	−30 03	9.5	4′	
W	Vr	18 05.0	−29 35	4.3–5.1	7.595 days	Cepheid
NGC 6546	OC	18 07.2	−23 20	8.0	13′	
NGC 6544	GC	18 07.3	−25 00	8.3	9′	
VX	Vr	18 08.1	−22 13	6.5–12.5	732 days	Semi-Regular

Sgr (continued)

Object	Typ	R.A. (2000) h m	Dec. (2000) ° '	Mag.	Size/Sep/ Period	Comments
Cr 367	OC	18 09.6	−23 59	6.4p	37'	
NGC 6568	OC	18 12.8	−21 36	8.6p	13'	
AP	Vr	18 13.0	−23 07	6.6–7.4	5.058 days	Cepheid
B92	**Dk**	**18 15.5**	**−18 11**		**12' × 6'**	**See M24**
M24	**OC**	**18 16.9**	**−18 29**	**4.5**	**90'**	**Small Sgr Star Cloud**
NGC 6595	OC	18 17.0	−19 53	7.0p	11'	
RS	Vr	18 17.6	−34 06	6.0–6.9p	2.416 days	Eclipsing Binary
NGC 6603	**OC**	**18 18.4**	**−18 25**	**11.1p**	**5'**	**In M24**
NGC 6613	**OC**	**18 19.9**	**−17 08**	**6.9**	**9'**	**M18**
NGC 6618	**DN**	**18 20.8**	**−16 11**	**7**	**46' × 37'**	**M17 Omega Nebula**
Y	Vr	18 21.4	−18 52	5.4–6.1	5.773 days	Cepheid
NGC 6624	GC	18 23.7	−30 22	8.3	6'	
NGC 6626	**GC**	**18 24.5**	**−24 52**	**6.9**	**11'**	**M28**
RV	Vr	18 27.9	−33 19	7.2–14.8	317.51 days	Long Period Variable
NGC 6638	**GC**	**18 30.9**	**−25 30**	**9.2**	**5'**	
NGC 6637	**GC**	**18 31.4**	**−32 21**	**7.7**	**7'**	**M69**
IC 4725	**OC**	**18 31.6**	**−19 15**	**4.6**	**32'**	**M25**
U	**Vr**	**18 31.9**	**−19 07**	**6.3–7.1**	**6.745 days**	**Cepheid (in M25)**
NGC 6642	GC	18 31.9	−23 29	8.8	5'	
V1017	Vr	18 32.1	−29 24	6.2–14.7p		Z And type?
NGC 6652	GC	18 35.8	−32 59	8.9	4'	
NGC 6656	**GC**	**18 36.4**	**−23 54**	**5.1**	**24'**	**M22**
NGC 6681	**GC**	**18 43.2**	**−32 18**	**8.1**	**8'**	**M70**
V350	Vr	18 45.3	−20 39	7.1–7.8	5.154 days	Cepheid
V356	Vr	18 47.9	−20 16	7.0–7.9p	8.896 days	Eclipsing Binary
YZ	Vr	18 49.5	−16 43	7.0–7.8	9.553 days	Cepheid
Ru 145	OC	18 50.5	−18 05		35'	
BB	**Vr**	**18 51.0**	**−20 18**	**6.7–7.3**	**6.637 days**	**Cepheid**
Cr 394	OC	18 53.5	−20 23	6.3p	22'	
NGC 6716	OC	18 54.6	−19 53	6.9	7'	
NGC 6715	GC	18 55.1	−30 29	7.7	9'	M54
NGC 6723	GC	18 59.6	−36 38	7.3	11'	
ST	Vr	19 01.5	−12 46	7.6–16.0	395.12 days	Long Period Variable
RY	Vr	19 16.5	−33 31	6.0–>15.0		Irregular; R CrB type
NGC 6774	OC	19 16.6	−16 16		30'	
R	Vr	19 16.7	−19 18	6.7–12.8	268.81 days	Long Period Variable
NGC 6809	GC	19 40.0	−30 58	7.0	19'	M55
54	**	19 40.7	−16 18	5.4,8.9	46"	42° (1932); 12767
NGC 6822	Gx	19 44.9	−14 48	9.2	10'	**Ir+** Barnard's Galaxy
V505	Vr	19 53.1	−14 36	6.5–7.5	1.183 days	Eclipsing Binary
RR	Vr	19 55.9	−29 11	5.6–14.0	334.58 days	Long Period Variable

(Continued on next page)

Sgr (continued)

Object	Typ	R.A. (2000) h m	Dec. (2000) ° '	Mag.	Size/Sep/ Period	Comments
RU	Vr	19 58.7	−41 51	6.0–13.8	240.31 days	Long Period Variable
NGC 6864	GC	20 06.1	−21 55	8.6	6'	M75
V1943	Vr	20 06.9	−27 13	9.0–11.0p		Irregular
RT	Vr	20 17.7	−39 07	6.0–14.1	305.31 days	Long Period Variable

Sagittarius is where the action is! In mythology, he was a centaur known for his skills as an archer. Our ancestors saw him keeping a watchful vigil over Scorpius to make certain that the scorpion did not misbehave. In today's sky, Sagittarius serves to mark the direction toward the center of our stellar megalopolis, the Milky Way.

To the unaided eye, Sagittarius bears little resemblance to a centaur, but is frequently likened to a teapot in profile. Observers in the northern hemisphere need a clear view to the south to see Sagittarius as it skims the southern horizon on warm summer nights. For amateurs close to and south of the Earth's equator, he blazes near the zenith, permitting an even more spectacular view of the treasures within (Figure 7.22). Indeed, there are so many wonderful deep sky objects scattered among the stars of the Archer that a whole book could be written just on them alone. Here, we highlight some of the more striking examples, as well as some of the lesser known sights.

M23 (NGC 6494) leads off our tour. Discovered by Messier in 1764, this is a rich open cluster of about 120 stars compressed into an area as large as the Full Moon. Most binoculars reveal a grainy texture, with a few of M23's brightest stars just breaking through a cloud of fainter suns. An unrelated 6th-magnitude star lies just beyond the cluster's northwest corner.

M20 (NGC 6514) is seen through most binoculars as a small, relatively faint puff of nebulosity about 4° north of the teapot's spout. These diffuse clouds can be rather difficult to discern through small instruments on hazy summer nights. Perhaps those were the conditions experienced by Messier when he first spotted M20. From his description, it seems that Messier may have never actually seen the tenuous interstellar clouds, but rather only the nebula's associated open cluster. William Herschel was the first to note that the nebulosity is cut into thirds by dark lanes. These apparent divisions have led to M20 being nicknamed the Trifid Nebula.

M8 (NGC 6523) is considered by most observers to be the premier diffuse nebula of the summer sky. On dark, moonless nights, it is

Figure 7.22
The Sagittarius-Ophiuchus border, highlighted by open cluster M23 (NGC 6494) at the upper left and the Snake Nebula (Barnard 72) at the lower right. Photograph by Martin Germano using a 135-mm lens at f/4 and a minus violet filter, hypered Technical Pan 2415 film, and a 30-minute exposure.

visible to the unaided eye as a hazy "island" amidst the "river" of our galaxy. Binoculars display a glowing cloud of great intricacy sliced in half by a dark lane, or "lagoon," of obscuring dust.

Messier described M8 as "a cluster which looks like a nebula in an ordinary telescope . . . but in a good instrument one observes only a large number of small stars." His account is especially interesting, since the open cluster **NGC 6530** lies in the foreground of M8. NGC 6530 is loosely comprised of about two dozen 7th- to 9th-magnitude stars and is a striking complement to the surrounding clouds. Barnard separately catalogued the regions of dark nebulae within the Lagoon Nebula as **Barnard 88, 89,** and **296** in his famous list of dark nebulosity.

M21 (NGC 6531) is another large, bright open cluster that is outstanding in binoculars. Found less than a degree northeast of the Trifid Nebula, it claims three dozen stars shining between magnitudes 8 and 12 within a span of 13′ of arc. If it were situated alone in the sky, M21 would still be a fine showpiece object. However, the added beauty of its magnificent surroundings moves this open cluster up the ranks to a "true stellar gem."

M24 is an unusual member of the Messier catalogue in many ways. For years, many deep sky object references identified it as NGC 6603, a small, tight open cluster in Sagittarius. However, this identification would seem to be in conflict with Messier's own description, which reads, "a large nebulosity in which there are many stars of different magnitudes . . . diameter 1° 30′." It was first suggested by Kenneth Glyn Jones of the British Astronomical Association that M24 is actually a larger region encompassing NGC 6603 and known as the Small Sagittarius Star Cloud. M24 is therefore easily visible to the naked eye whenever the summer Milky Way is present. Look for a brighter region along the mainstream of our galaxy south of the more prominent Scutum Star Cloud. Binoculars reveal a myriad of stars strewn throughout the region, just as Messier described.

NGC 6603 must now be considered a separate object. In 7× binoculars, it is just visible as a dim, unresolved glow slightly northeast of M24's center. With stars no brighter than about 10th magnitude, NGC 6603 must be viewed through the largest binoculars before resolution can be achieved.

Located on the northwest corner of M24 is the dark nebula **Barnard 92**. Set against a magnificent field of distant suns, this cloud may be glimpsed with 7× and grows quite apparent with 11× and above. Look for a black "hole" (*not* a real black hole!) measuring about one-fifth of our Moon's diameter across.

M18 (NGC 6613) is one of the less impressive Messier objects. Consisting of about 18 loosely packed stars, it appears in binoculars as a fairly obvious stellar arrangement. The four brightest stars in this open cluster are set in a triangular pattern and may just be glimpsed in 10× glasses. The rest of its stars blend into a faint haze.

M17 (NGC 6618) is found north of M18 along the summer Milky Way. The Swiss astronomer De Cheseaux discovered this stunning region of bright nebulosity in the spring of 1764, beating Messier by only a few months. Messier later described it as "a train of light without stars . . . in the shape of a spindle." The famed observer William Herschel was the first to liken this nebula's form to the Greek letter omega; ergo its modern name, the Omega Nebula.

The Omega Nebula is a spectacular sight through nearly all binoculars. Giant glasses reveal an intricate cloud in the shape of an extended number "2." In fact, the cloud's long, curved "neck" has led some to dub it the Swan Nebula. Adding a nebula filter creates an interstellar spectacle whose magic cannot be conveyed in words or pictures.

M28 (NGC 6626) was first discovered in 1764 by Messier, who identified it as a "nebula containing no star." Through today's binoculars, M28 is also seen as only a small smudge of faint light just northwest of Lambda Sagittarii. Although its true nature remains a mystery through low-power glasses, M28 is famous as a fine globular cluster. The brightest of its 100,000 stars only reach about 12th magnitude, well below the limit of conventional binoculars. The distance to M28 is believed to be 20,000 light years.

M69 (NGC 6637) is the westernmost of three Messier globulars that reside along the "bottom" of the teapot (the others are M70 and M54). Of 8th magnitude, it is found about 2½° northeast of Epsilon Sagittarii, and directly adjacent to an 8th-magnitude field star. Lacaille discovered M69 in 1752. When Messier saw it eight years later, he described it as "a nebula without a star . . . near a star." This is an accurate, though somewhat sterile, description of the object's appearance in binoculars. The stars of M69 require a moderately large telescope to be seen, as none shine brighter than about 14th magnitude.

M25 (IC 4725) is a bright open cluster that is easily found through binoculars about 6° north of the teapot's lid. From our earthly vantage point, the group subtends ½° and is thought to lie about 1,900 light years away.

M25 hosts 30 stars in all, half of which are bright enough to be seen individually through 7× glasses. The cluster is unique in two respects. First, it is the only Messier object to be cross-listed in the *Index Catalogue* (IC). Second, it is one of a handful of open clusters known to contain a Cepheid-type variable star: **U Sagittarii** fluctuates between magnitudes 6.3 and 7.1 over a period of about 6 days, 18 hours. Although the variations are quite subtle, a trained eye can estimate the star's period and apparent luminosity quite accurately.

M22 (NGC 6656) is one of the most spectacular globular clusters in the heavens, as evidenced in Figure 7.23. Under ideal conditions, it is visible to the unaided eye as a hazy spot northeast of Lambda Sagittarii. Seven-power binoculars readily show its nebulous disk, while giant glasses are just able to resolve a couple of individual stars against a grainy surface.

Were it not for its low altitude from the northern hemisphere, M22 might be the standard by which all other globulars are judged, rather than better known M13 in Hercules. In fact, photographs show that M22 spans 24' of arc, about 50% more than the Hercules cluster spans. Apparently, the reason for this is M22's closer distance to Earth. As studies indicate, it probably contains less than half as many

Figure 7.23
M22 (NGC 6656) in Sagittarius. Photograph by Lee Coombs using a 10-inch f/5 Newtonian reflector and a 15-minute exposure on 103a-E spectroscopic film.

stars as the Hercules cluster. M22, believed to be about 10,000 light years from Earth, has a real diameter of about 50 light years, while M13 is about 24,000 light years away and some 170 light years across.

M70 (NGC 6681) is the second of three Messier globular clusters found between Epsilon and Zeta Sagittarii. In many ways, M70 and M69 are twin objects. Both are within 2½° of each other, shine at close to 8th magnitude, and appear about 8 arc-minutes across. Recently completed investigations conclude that the two even lie relatively close to each other in space, with M70 about 35,000 light years from Earth and M69 approximately 33,500 light years away.

BB Sagittarii is a bright Cepheid variable seen to the east of 29 Sagittarii. Looking like a golden ember, it varies only slightly between magnitudes 6.7 and 7.3 with a period of just over 6.5 days. BB forms a bright optical double star with an unrelated 6th-magnitude star to its east.

Just to the northeast of BB Sagittarii lie the weak open clusters **Collinder 394** and **NGC 6716**. Collinder 394 is a feeble spray of stars placed just east of the variable. Look for a small, ill-defined asterism of five magnitude 9 suns. Just to its northeast, and well within the same binocular field, is NGC 6716. This group is composed of 20 stars, although only four are as bright as 10th magnitude. The remaining stars are faintly held within the cluster's 7'-of-arc boundaries. Observers should note that a number of brighter suns surround NGC 6716 and tend to obscure exactly where the cluster begins and ends.

M54 (NGC 6715) is the easternmost of the three Messier globulars along the bottom of the teapot. Binoculars reveal its 8th-magnitude disk set among a field rich in stars about 2° west of Zeta Sagittarii. On the 12-point globular cluster concentration scale (with I indicating the highest concentration) set up by Harlow Shapley and Helen Hogg in 1927, densely packed M54 rates a III. This makes it one of the densest objects of its kind in the Messier catalogue. At a projected distance of 70,000 light years, M54 is also the most distant globular cluster in the Messier catalogue.

NGC 6774 is a weak open cluster positioned about 1½° west of 5th-magnitude Nu Sagittarii. Many observing handbooks pass by NGC 6774 in favor of other objects in Sagittarius, perhaps because the cluster's ½° extent makes it difficult to identify through telescopes. Binoculars, however, readily display about 20 scattered stars; look for a triangular asterism of 7th-magnitude stars with many fainter suns set within.

M55 (NGC 6809) is the most spectacular deep sky object in eastern Sagittarius. Lacaille discovered this giant globe of stars in 1752 while observing from the Cape of Good Hope. Messier saw it for the first time 26 years later, when he described it as "a nebula which is a whitish spot; extending for 6' around, the light is even and does not appear to contain a star." Through 7× binoculars, we see that Messier was mistaken: M55 does contain a star—one, to be exact! A lone 9th-magnitude sun is the brightest of the cluster's bevy of stars and the only one seen in 10× glasses. With 15×, several points of light are tentatively visible within M55, but averted vision will probably be needed to confirm their existence. The actual number of stars within this loosely structured globular cluster is probably between 75,000 and 100,000.

54 Sagittarii is an easily separable double star through binoculars. The colorful pair collectively forms the southwestern member of a curved stellar asterism of three 5th-magnitude stars found about 9° due west of Beta Capricorni. The components of 54 Sagittarii are magnitudes 5.4 and 8.9 and have a separation of about 46" of arc. The spectral classes of the two stars are type-K and type-F, respectively, forming a colorful contrast between the orangish primary and white secondary.

NGC 6822 is a challenging object for observers even with the largest binoculars. Most amateurs know it as "Barnard's Galaxy," after its 1884 discoverer, E. E. Barnard. Due to the galaxy's large apparent diameter, the resulting low surface brightness effectively masks its existence among the stars of eastern Sagittarius. It is most easily located by first zeroing in on previously mentioned 54 Sagittarii and then sliding just a bit northeast to 55 Sagittarii. Once at 55, mentally note the span between it and another magnitude 5 star to the northeast. NGC 6822 lies northeast of this anonymous sun by the same distance that 55 lies to its southwest. Look for an irregular, oval glow with a major axis running north to south. Photographs reveal what is suspected visually—an irregular galaxy similar in appearance to the Magellanic Clouds. At an estimated distance of about 1.7 million light years, NGC 6822 is well within the limits of the Local Group of galaxies, which also includes the Milky Way, the Andromeda Galaxy, and M33 in Triangulum.

M75 (NGC 6864), at an estimated distance of 59,000 light years, is second only to M54 as the most remote globular cluster catalogued by Messier. Visually, it is a bright, compact object appearing as a perfectly round glow surrounding a brighter nucleus. With an apparent diameter of only 6′ of arc, M75 may require some searching before it is found, especially through lower power glasses.

Scorpius (Sco)

Object	Typ	R.A. (2000) h m	Dec. (2000) ° ′	Mag.	Size/Sep/ Period	Comments
Nu	**	16 12.0	−19 28	4.3,6.4	41″	337° (1955); 9951
NGC 6093	GC	16 17.0	−22 59	7.2	9′	M80
B41,43	Dk	16 22	−19 40		200′ × 80′	
NGC 6121	GC	16 23.6	−26 32	6.0	26′	M4
NGC 6124	OC	16 25.6	−40 40	5.8	29′	
NGC 6178	OC	16 35.7	−45 38	7.2	4′	
NGC 6231	OC	16 54.0	−41 48	2.6	15′	
B1833	**	16 54.0	−41 48	5.6,7.3	57″	21° (1847); in NGC 6231
Cr 316	OC	16 55.5	−40 50	3.4p	105′	
NGC 6242	OC	16 55.6	−39 30	6.4	9′	
RS	Vr	16 55.6	−45 06	6.2–13.0	320.06 days	Long Period Variable
RR	Vr	16 56.6	−30 35	5.0–12.4	279.42 days	Long Period Variable
V861	Vr	16 56.6	−40 49	6.1–6.7	7.848 days	Eclipsing Binary; in Tr 24
Tr 24	OC	16 57.0	−40 40	8.6p	60′	
NGC 6249	OC	16 57.6	−44 47	8.2	6′	
RV	Vr	16 58.3	−33 37	6.6–7.5	6.061 days	Cepheid
RT	Vr	17 03.5	−36 55	7.0–16.0	449.04 days	Long Period Variable
HD 153919	*	17 03.9	−37 51	6.6		X-ray source
NGC 6281	OC	17 04.8	−37 54	5.4	8′	
AH	Vr	17 11.3	−32 30	8.1–12.0p	713.6 days	Semi-Regular
NGC 6322	OC	17 18.5	−42 57	6.0	10′	
V636	Vr	17 22.8	−45 37	6.0–6.9	6.797 days	Cepheid
Harvard 16	OC	17 31.4	−36 51		15′	
NGC 6383	OC	17 34.8	−32 34	5.5	5′	
Tr 27	OC	17 36.2	−33 29	6.7	7′	
NGC 6388	GC	17 36.3	−44 44	6.9	9′	
Tr 28	OC	17 36.8	−32 29	7.7	8′	
Cr 338	OC	17 38.2	−37 34	8.0p	25′	
NGC 6405	OC	17 40.1	−32 13	4.2	15′	M6, Butterfly Cluster
NGC 6400	OC	17 40.8	−36 57	8.8p	8′	
BM	Vr	17 41.0	−32 13	6.8–8.7p	850 days	Semi-Regular; in M6
Tr 29	OC	17 41.6	−40 06	7.5p	9′	
RU	Vr	17 42.4	−43 45	7.8–13.7	369.20 days	Long Period Variable
NGC 6416	OC	17 44.4	−32 21	5.7	18′	
NGC 6441	GC	17 50.2	−37 03	7.4	8′	

(Continued on next page)

Sco (continued)

Object	Typ	R.A. (2000) h m	Dec. (2000) ° '	Mag.	Size/Sep/ Period	Comments
RY	Vr	17 50.9	−33 42	7.5–8.4	20.316 days	Cepheid
NGC 6453	GC	17 50.9	−34 36	9.9	4'	
NGC 6475	OC	17 53.9	−34 49	3.3	80'	M7
NGC 6496	GC	17 59.0	−44 16	9.2	7'	

With its long, winding body deeply immersed in the plane of our galaxy, **Scorpius,** the Scorpion, is a veritable playground for the binocularist. Within its borders is a bountiful selection of beautiful deep sky objects strewn across some of the finest heavenly star fields. One of my favorite ways to spend a warm, clear summer evening is to simply sit back and casually scan the Scorpion's crooked body from head to toe. Along the way are many elegant asterisms, clusters, and nebulae that provide hours of fascinating sky watching.

M80 (NGC 6093) is a bright globular cluster found about halfway between brilliant Antares (Alpha Scorpii) and Graffias (Beta Scorpii). History is not clear as to who discovered M80, for its first sighting was made almost simultaneously by Messier and Méchain in January 1781.

Binoculars readily detect the 7th-magnitude disk of M80 spanning 9' of arc. Set in a striking star field, it presents itself as a softly glowing, round blur highlighted by a more intense center. Resolution of the stars in M80 requires at least an 8- to 10-inch telescope, as the brightest stellar elements are 14th magnitude.

M80 has the distinction of being one of the few globular clusters ever to spawn a nova. The event was seen in May 1860, when an anonymous star suddenly flared to 7th magnitude. Now known as T Scorpii, the star slowly faded back into obscurity in about three months.

M4 (NGC 6121) is an exciting globular cluster to view through just about all optical instruments. Some sharp-eyed observers have reported seeing this 6th-magnitude object without any optical aid. Binoculars make it plainly visible as a large, amorphous globe only 1½° west of Antares. Visual observers typically note M4 as only about half as large as its 26-arc-second photographic extent.

The brightest stars in this distant horde just crack the 11th-magnitude barrier. Their brightness, combined with the cluster's fairly loose structure, makes partial resolution possible in 15× glasses under ideal sky conditions. Perhaps even easier to see is a bright "bar" of light slicing across the cluster's center. This unusual feature is unmistakable in giant glasses and even perceptible in steadily held 7×50's. At nearly 7,000 light years away, M4 is considered by some authorities to be the closest globular cluster to Earth. (Others believe that NGC 6397 in Ara is slightly closer.)

NGC 6124 is a fine open cluster lying close to the Scorpius-Norma-Lupus border. Binoculars display a rich group of about two dozen 9th-magnitude points of light covering an area equal to the Moon's. A distinguishing chain of seven stars runs along the cluster's southern edge, while a closely packed group of five luminaries highlights its center.

NGC 6231 lies between Mu^{1+2} Scorpii and Zeta^{1+2} Scorpii, one of the most remarkable regions in the entire sky. The cluster itself is a tight knot of about 120 stars crushed into 15' of arc. Many of its intrinsically brilliant, hot type-O and type-B suns shine between 6th and 8th magnitude and are easily seen in all binoculars.

NGC 6231 is thought to be about 5,900 light years away. If we could reduce that distance to about 400 light years (the same distance as the Pleiades), then the brightest stars of NGC 6231 would rival Sirius in appearance! One of the cluster's brightest members is the binary star **van den Bos 1833** (**B1833**), which pairs a magnitude 5.6 principal star with a magnitude 7.3 companion.

The entire region between and around Mu and Zeta Scorpii is a dazzling tract of sky. Stars of all colors and contrasts against an inky black sky are strewn across a binoculars' field of view. Many of us have become so accustomed to looking for specific objects that we tend to ignore the larger scene. But not here. Pause for a moment, scan slowly through wide-field glasses, and absorb the awesome beauty of the heavens. After all, that *is* why we are amateur astronomers.

Collinder 316 and **Trumpler 24** contain many of the stars between Zeta and Mu Scorpii. The former open cluster is a large, rich stellar accumulation that spreads across nearly 2°. Its many stars range from 6th to less than 9th magnitude and offer a scenic view through 7× glasses.

Trumpler 24, also identified as Harvard 12 in some references, is partially superimposed on the eastern edge of Collinder 316. Trumpler 24 is only about half as large as its intruding neighbor and is characterized by an arc of three 6th-magnitude stars extending north-south. The southernmost of these stars is the eclipsing binary **V861 Scorpii.** The unseen companion of V861 orbits the primary in 7.848 days, causing an apparent shift in brightness between magnitudes 6.1 and 6.7.

NGC 6242 lies between Collinder 316 and Mu Scorpii. Binoculars resolve a lone 6th-magnitude sun attended by four or five 9th-magnitude points of light and the subtle glow of fainter, invisible cluster members. In all, approximately 45 stars down to 11th magnitude compose NGC 6242.

NGC 6281 may be found about one-third of the way between Mu Scorpii and Nu Scorpii, the Scorpion's stinger. Here, 7× glasses find about half a dozen stars between magnitudes 7 and 9 set in a crooked cruciform.

Just beyond the western edge of NGC 6281 is an innocent looking 6th-magnitude star that has been identified as an X-ray source. Catalogued as **HD 153919,** it emits about 800 times the X-radiation as our Sun and is thought to be an eclipsing binary stellar system.

NGC 6322 lies between Eta and Theta Scorpii along the southernmost part of the Scorpion's tail. Binoculars show only about five stars, as well as a faintly perceptible glow from the combined light of an additional 25 suns that belong to this open cluster.

M6 (NGC 6405) is one of two large, bright open clusters found north of the Scorpion's tail. (The other, M7, is discussed shortly.) The naked eye is all that is needed to detect M6 as a dim smudge of light against the glow of the Milky Way's galactic plane. Through binoculars, M6 lives up to its Butterfly Cluster nickname, as two "wings" of stars spread out from the cluster's more densely packed center (Figure 7.24). Seven-power glasses resolve about 30 stars across the 15'-of-arc face of M6, with 11× binoculars revealing an additional dozen faint points of light. In all, 80 stars belong to this cluster.

The brightest star in M6 is an orange stellar ember found east of the cluster's center. This star is **BM Scorpii,** a semi-regular variable whose brightness fluctuates between magnitudes 6.8 and 8.7. It takes 850 days, on average, for the star to complete one cycle.

M7 (NGC 6475) is the brightest, and one of the most dazzling, open clusters in all of Scorpius. It is easily visible to the unaided eye as a misty glow about 4° northeast of Lambda Scorpii, the star marking

Figure 7.24
Portions of Sagittarius, Scorpius, and Ophiuchus contribute to this wide-field view aimed toward the center of the Milky Way. Note M6 (NGC 6405) and M7 (NGC 6475) to the lower right and M8 (NGC 6523) above center. Photograph by Jim Barclay, F.R.A.S.

the Scorpion's stinger. Records show that M7 was known long before the invention of the telescope: the second-century astronomer Ptolemy cited it as a nebulous patch in his monumental work, the *Almagest*.

When viewed with even a pair of modest field glasses, M7 bursts into an exceptionally beautiful array of stars spanning 80 arc-minutes. Of the 80 stars identified as cluster members, more than 30 are brighter than 10th magnitude and are visible in binoculars. The brightest star of M7, a type-G sun of 6th magnitude, gleams close to the group's center. Many fainter cluster stars also shine with a yellowish cast, while others appear blue-white. Measurements indicate that M7 is approximately 260 billion years old, which is nearly twice as old as many other youthful open clusters, including neighboring M6.

Sculptor (Scl)

Object	Typ	R.A. (2000) h m	Dec. (2000) ° '	Mag.	Size/Sep/ Period	Comments
Blanco 1	OC	00 04.3	−29 56	4.5	90′	Zeta Scl cluster
NGC 55	Gx	00 14.9	−39 11	8p	32′ × 6′	SBm
S	Vr	00 15.4	−32 03	5.5–13.6	365.32 days	Long Period Variable
NGC 253	Gx	00 47.6	−25 17	7.1	22′ × 6′	Scp
NGC 288	GC	00 52.8	−26 35	8.1	14′	
Y	Vr	23 09.1	−30 08	8.7–10.3p		Semi-Regular
NGC 7793	Gx	23 57.8	−32 25	9.1	9′ × 7′	Sdm

Sculptor, devised by Lacaille to represent a sculptor's studio, is formed from a set of faint stars to the east of the bright autumn star Fomalhaut in Piscis Austrinus and south of Beta Ceti in Cetus. Sculptor is best known as the home of a group of galaxies called the "Sculptor Group." At about 12 million light years away, this galactic gathering is one of the closest collections to the Milky Way's Local Group.

Blanco 1, an ill-defined open cluster in central Sculptor, is much closer to home than the galaxies for which the constellation is famous. Even though the 30 stars in this cluster are set against an empty region close to the south galactic pole, they give little indication of forming an open cluster. Binoculars show about a dozen stars brighter than 8th magnitude, as well as a smattering of fainter suns, all within the group's 90′-of-arc span. Highlighting Blanco 1 and acting as a beacon to show us the way is 5th-magnitude Zeta Sculptoris.

Figure 7.25
The spectacular spiral galaxy NGC 253 in Sculptor. Photograph by Lee Coombs using a 10-inch f/5 Newtonian reflector and a 15-minute exposure on 103a-O spectroscopic film.

NGC 55 is a large, edge-on barred spiral in extreme southern Sculptor. Unfortunately, the galaxy's southern location causes most observers in the mid-northern hemisphere to miss out on its true splendor. From a southern vantage point, NGC 55's long, needle-thin disk shines with a uniform gray light through 7× glasses, while giant binoculars begin to suggest a grainy texture to its surface. Photographs reveal this unusual mottled appearance stretching across the full ½° extent of NGC 55.

NGC 253 (Figure 7.25) is said by many observers to be the finest spiral galaxy found south of the celestial equator. Nestled gently within an asterism of half a dozen 7th-magnitude suns about 7° due south of Beta Ceti, NGC 253 is a long, cigar-shaped splinter set off by a bright central core. Under extraordinary sky conditions, a hint of surface texture can be glimpsed through 11×80 and larger instruments. Both NGC 55 and NGC 253 are members of the Sculptor Group of galaxies.

NGC 288 lies about 2° southeast of NGC 253. Shining at 8th magnitude and spanning 14′ of arc, this globular cluster is seen in binoculars as a hazy ball of cotton framed by several faint field stars. NGC 288 is famous for its loose stellar structure. Large-aperture 20× glasses just begin to reach some of the myriad of stars that constitute this distant horde.

Scutum (Sct)

Object	Typ	R.A. (2000) h m	Dec. (2000) ° '	Mag.	Size/Sep/ Period	Comments
B95	Dk	18 25.6	−11 45		30'	2.6° NE of M16
RZ	Vr	18 26.6	−09 12	7.3–8.8	15.190 days	Eclipsing Binary
B97	Dk	18 29.1	−09 56		50'	1° NW of NGC 6649
B312	Dk	18 30.9	−15 08		100' × 30'	2.5° E of M17
Do 29	OC	18 31.4	−06 38		18'	Asterism?
B100–1	Dk	18 32.7	−09 08		40' × 15'	1.4° N of NGC 6649
Do 30	OC	18 32.9	−06 02		18'	Asterism?
NGC 6664	OC	18 36.7	−08 13	7.8	16'	
B314	Dk	18 37.7	−09 37		35' × 25'	1° NE of NGC 6649
B103	Dk	18 39.2	−06 37		40'	NW of Sct Star Cld
NGC 6682	OC	18 41.6	−04 46			Star field
NGC 6683	OC	18 42.2	−06 17	9.5p	11'	Star field
NGC 6694	OC	18 45.2	−09 24	8.0	15'	M26
B104	Dk	18 47.3	−04 32		16' × 1'	20' N Beta Sct
R	Vr	18 47.5	−05 42	4.5–8.2	140.05 days	RV Tauri type
B108	Dk	18 49.6	−06 19		3'	0.5° W of M11
B318	Dk	18 49.7	−06 24		90' × 2'	S of M11
B111,119a	Dk	18 51	−05		120' × 120'	Two crescents N of M11
NGC 6705	OC	18 51.1	−06 16	5.8	14'	M11; Wild Duck Cluster
B112	Dk	18 51.2	−06 40		20'	S of M11
NGC 6712	GC	18 53.1	−08 42	8.2	7'	
B114–8	Dk	18 53.2	−07 06		50' × 5'	SE of M11

Scutum, the Shield, is one of the most recently appointed constellations of the summer sky, having been created by Hevelius in 1690. Scutum is said to represent the shield of John Sobrieski III of Poland, who successfully resisted the Turkish advance on Vienna in 1683.

The brightest stars of Scutum are only 3rd magnitude and are easily lost in the glory of the Scutum Star Cloud (Figure 7.26). This region is among the most beautiful to scan with binoculars. Innumerable stars are sprinkled throughout Scutum with a richness that is unsurpassed in the northern hemisphere. E. E. Barnard referred to the

Figure 7.26
The Scutum Star Cloud, as photographed by George Viscome. Exposure time was 30 minutes on ISO 400 film through a 135-mm lens.

Scutum Star Cloud as "the gem of the Milky Way," a title that only begins to convey the magnificence of this part of the summer sky.

Barnard 103 is one of many patches of dark nebulosity within Scutum. The irregular boundaries of this obscuring cloud cover 40' of arc and appear almost amoeba-shaped in photographs. Nary a star can be seen across its face through binoculars.

M26 (NGC 6694), located less than a degree southeast of 5th-magnitude Delta Scuti, is a small, condensed open cluster. Its 30 stars, none of which shines brighter than magnitude 10.3, combine to produce an 8th-magnitude misty glow spanning 15' of arc.

R Scuti, easily found 1° due south of Beta Scuti, exhibits characteristics of the unusual RV Tauri family of variable stars. Across a varying period that averages 140 days, this yellowish star flickers between 5th and 6th magnitude, although it can reach magnitude 4.5 and drop to less than 8th. A much longer secondary cycle of about 1,300 days has also been detected.

M11 (NGC 6705), famous for its high density, as seen in Figure 7.27, lies between Earth and the northern boundary of the Scutum Star Cloud. Discoverer Gottfried Kirch noted in 1681 that M11 seemed like "a small, obscure spot with a star shining through." The famed 18th-century observer Admiral Smyth noted that M11 resembles "a

Figure 7.27
The Wild Duck cluster, M11 (NGC 6705), as photographed by George Viscome. He used a 3-inch f/6.6 refracting telescope for this 20-minute exposure on ISO 400 film.

flight of wild ducks." With this, M11's nickname of the Wild Duck Cluster was born.

Through binoculars, M11 looks more like an unresolved globular cluster than an open cluster, although it is definitely the latter. Its hundreds of stars all shine between 11th and 14th magnitude, except for Kirch's lone 8th-magnitude maverick. Though the "ducks" are a bit too "wild" to be resolved through most glasses, the cluster's characteristic arrowhead shape may be inferred.

Of the many dark nebulae that surround M11, **Barnard 111** and **Barnard 119a** are the most obvious. Visually, both appear as large, kidney-shaped patches that almost touch each other at their southern tips. Although they visually impress me as being nearly identical in size, Barnard 111 proves to be about twice as large as Barnard 119a in photographs. Adding to the area's attractiveness is a "peninsula" of about half a dozen magnitude 8 to magnitude 10 stars wedged between the nebulae.

Binoculars reveal **Barnard 318** as a pencil-thin black cloud stretching east to west just south of M11. A bit farther south is **Barnard 112,** an egg-shaped dark nebula measuring about 20′ across. As with all dark nebulae, transparent skies and sharp eyes are needed to spot these elusive objects.

Serpens Cauda and Serpens Caput (Ser)

Object	Typ	R.A. (2000) h m	Dec. (2000) ° ′	Mag.	Size/Sep/ Period	Comments
NGC 5904	GC	15 18.6	+02 05	5.8	17′	M5
S	Vr	15 21.7	+14 19	7.0–14.1	368.59 days	Long Period Variable
Tau[4]	Vr	15 36.5	+15 06	7.5–8.9		Irregular
R	Vr	15 50.7	+15 08	5.2–14.4	356.41 days	Long Period Variable
U	Vr	16 07.3	+09 56	7.8–14.7	237.85 days	Long Period Variable
Nu	**	17 20.8	−12 51	4.3,8.3	46″	28° (1959); 10481
NGC 6539	GC	18 04.8	−07 35	9.6	7′	
NGC 6605	OC	18 17.1	−14 58	6.0p		
NGC 6604	OC	18 18.1	−12 14	6.5	2′	
NGC 6611	DN/ OC	18 18.8	−13 47	6.0	35′	**M16, Eagle Nebula**
59	Vr	18 27.2	+00 12	4.9–5.9		Unknown type
IC 4756	OC	18 39.0	+05 27	5.4p	52′	
Theta	**	18 56.2	+04 12	4,5,8	22″,414″	104° (1973), 56° (1927); 11853

Serpens, the Serpent, has the unique characteristic of being the only constellation broken in half by a second star pattern (Ophiuchus). To the west is Serpens Caput, the head. This is a sparse area highlighted by a fine globular cluster and several variable stars within range of binoculars. East of Ophiuchus is Serpens Cauda, the tail. Even with the galactic plane passing through, eastern Serpens contains surprisingly few deep sky objects. But what is lacking in quantity is more than made up for in quality.

M5 (NGC 5904) is a highlight of Serpens Caput and one of the finest globular clusters located north of the celestial equator. It was first spied by Kirch and later added to Messier's famous list of comet look-alikes. Believed to contain over half a million stars, M5 is bright enough to be seen without any optical aid on dark, clear nights. It is readily apparent as a nebulous globe through just about all binoculars. Giant glasses show the disk rapidly brightening toward the center, with perhaps the slightest hint of a mottled surface. With steady seeing, the first few individual stars are just suspectible in $25\times$ to $30\times$ binoculars.

M16 (NGC 6611), in Serpens Cauda, has come to be known to visual observers as a fine open cluster of about five dozen suns scattered across half a degree of sky. Seven-power binoculars show the dozen brightest suns, along with the unmistakable glow of fainter suns. Deep photographs discover a great complex of beautiful nebulosity threaded among the cluster stars. Until recently, the nebula, known popularly as the Eagle Nebula, was considered a difficult visual challenge. However, thanks to the widespread use of contrast-enhancing nebula filters, the Eagle may now be seen to soar where it never had been seen before. Through 11×80 binoculars paired with nebula filters, M16 is a bright smattering of stars engulfed in wonderful bright and dark clouds.

IC 4756, with an apparent diameter of nearly 1°, is a bright group of stars found in Serpens Cauda. Eighty luminaries of magnitude 7 and fainter are nestled within this open cluster and form an impressive sight against the brilliant backdrop of the summer Milky Way. Note that as magnification increases, the dramatic clustering effect is quickly lost. Therefore, $7\times$ to $9\times$ glasses prove to be the best instruments for viewing a cluster like IC 4756.

Sextans (Sex)

Object	Typ	R.A. (2000) h m	Dec. (2000) ° '	Mag.	Size/Sep/ Period	Comments
NGC 3115	Gx	10 05.2	−07 43	9.1	8' × 3'	E6 Spindle Galaxy
RT	Vr	10 12.3	−10 19	8.0–8.5	96 days	Semi-Regular

Sextans was created by the 17th-century astronomer Hevelius to honor the instruments that he used to measure the positions of the stars. The celestial Sextant is home to many galaxies, but all fall below binocular visibility save one. To hold our interest a bit longer, there is also a semi-regular variable star within binocular range.

NGC 3115, nicknamed the Spindle Galaxy for its distinctive lens shape, is an outstanding example of an E6 galaxy. Through 7× glasses, it appears as an amorphous, ill-defined cloud, while 11× binoculars begin to hint at its unique shape.

While NGC 3115 is well known to deep sky aficionados, few casual observers take the time to locate this superb object due to its rather bleak surroundings. The easiest way to find NGC 3115 is to locate Alpha Hydrae (Alphard) first. Then scan about 11° due east until you come upon the wide optical double star 17 and 18 Sextanis. The galaxy lies about 1½° to the northeast of this duo.

Taurus (Tau)

Object	Typ	R.A. (2000) h m	Dec. (2000) ° '	Mag.	Size/Sep/ Period	Comments
21 + 22	**	03 46.1	+24 32	5.6,6.4	168"	Asterope
NGC 1435	DN	03 46.1	+23 47		30' × 30'	Temple's Nebula (Merope)
M45	OC	03 47.0	+24 07	1.2	110'	Pleiades
Eta	**	03 47.5	+24 06	3,8,8,8	117",181",191"	289°, 312°, 295° (1903); Alcyone
27 + BU	**	03 49.2	+24 03	3.7,5.0	300"	180°; Atlas & Pleione
BU	Vr	03 49.2	+24 08	4.8–5.5		Irregular; Pleione
H VI 98	**	04 15.5	+06 11	6.3,7.0	66"	315° (1937); 3085
Phi	**	04 20.4	+27 21	5.0,8.4	52"	250° (1925); 3137
Kappa	**	04 25.4	+22 18	4.4,5.4	340"	173° (1923)
Mel 25	OC	04 27	+16	0.5	330'	Hyades
Theta^{1+2}	**	04 28.7	+15 52	3.8,3.4	337"	346° (1921)

(Continued on next page)

Tau (continued)

Object	Typ	R.A. (2000) h m	Dec. (2000) ° '	Mag.	Size/Sep/ Period	Comments
88	**	04 35.7	+10 10	4.3,8.4	70″	299° (1920); 3317
HU	Vr	04 38.3	+20 41	5.9–6.7	2.056 days	Eclipsing Binary
NGC 1647	OC	04 46.0	+19 04	6.3	45′	
NGC 1746	OC	05 03.6	+23 49	6.0	45′	
NGC 1807	OC	05 10.7	+16 32	7.0	17′	
NGC 1817	OC	05 12.1	+16 42	7.7	15′	
CD	Vr	05 17.5	+20 08	7.3–7.9	3.435 days	Eclipsing Binary
DoDz 3	OC	05 33.7	+26 29		15′	
NGC 1952	DN	05 34.5	+22 01	8.2	6′ × 4′	M1 Crab nebula (SNR)
DoDz 4	OC	05 35.9	+25 27		28′	
OΣΣ 67	**	05 48.4	+20 52	6.1,8.6	76″	161° (1933); 4392

Standing proudly among the stars of the northern hemisphere's winter sky is the fiery celestial Bull, **Taurus**. Most artists depict him as daring nearby Orion, the Hunter, into action. The many fine deep sky objects within the Bull also dare amateurs into action even on the coldest winter nights. Taurus embraces many eye-catching double and variable stars, some striking binocular open clusters, and even a 6,000-year-old stellar corpse.

M45 (Figure 7.28), better known as the Pleiades or Seven Sisters, is one of the sky's premier open clusters. Most naked-eye stargazers can count six or seven stars here set in the shape of a tiny dipper, although many more are visible under superior sky conditions. The modern record was set by veteran deep sky observer Walter Scott Houston from Tucson, Arizona, in 1935, when he glimpsed 18 Pleiads!

The Pleiades cluster is composed of at least 100 stars scattered across 110′ of arc—nearly four times the size of the Full Moon. While this is too large to fit comfortably into most telescopic fields of view, binoculars provide a spectacular scene. I can recall the outstanding view I had of the Pleiades on a recent frigid but crystalline night; through a pair of 11×80 glasses, the scene nearly left me breathless! Across the field were strewn dozens of stellar diamonds and sapphires.

Several striking pairs of double and multiple stars highlight M45. **Atlas** (27 Tauri) and **Pleione** (28 Tauri) form a wide, optical binary system. Atlas shines at magnitude 3.7, while five arc-minutes away, Pleione (also known as **BU Tauri**) varies irregularly between magnitudes 4.8 and 5.5. Together, they form the eastern-pointing "handle" of the Pleiades' "bowl."

Other double stars within the Pleiades that are detectable

Figure 7.28
The dazzling open cluster M45 (the Seven Sisters) in Taurus. Photograph by Johnny Horne using an 8-inch f/1.5 Schmidt camera, hypered Technical Pan 2415 film, and a six-minute exposure.

through binoculars include **Asterope** (21 and 22 Tauri), a wide pair of stars shining at magnitude 5.6 and 6.4 and separated by nearly 3' of arc. **Alcyone** (Eta Tauri), the brightest Pleiad, is an easily resolvable quadruple star system: the brilliant magnitude 2.9 primary is attended by a trio of 8th-magnitude points of light.

If you have an exceptionally clear evening, look closely and you just might glimpse soft gossamer wisps surrounding some of the brighter cluster stars. These gentle clouds are all that remain of a once-mammoth nebula from which the Pleiades were formed about 78 million years ago. The brightest portion of the Pleiades nebulosity is **NGC 1435,** found around Merope, the southeastern star in the Pleiades' "bowl." W. Tempel discovered this comet-shaped cloud in October 1859. He described its tenuous nature as resembling "a breath on a mirror." Incidentally, the stars of the Pleiades appear so bright through binoculars that many amateurs mistake the glow of scattered starlight for the cluster's nebulosity. While scatter can never be completely eliminated, it is greatly reduced if all optical surfaces are clean and free of contamination. Otherwise, with dirty optics, clouds will be seen around every star in the sky!

Melotte 25 is the catalogue entry for another old friend, the Hyades star cluster. The Hyades are well known as the V-shaped group of naked eye stars that make up the head of Taurus, but there is much more to this cluster than meets the eye. Hundreds of stars are contained within the group's 5½° span, with 132 of them brighter than 9th magnitude and, therefore, visible in binoculars. In low-power, wide-field glasses, the Hyades ranks as another of the northern winter sky's best clusters.

The brightest star in the Hyades is **Theta^2 Tauri**, at magnitude 3.4. It teams with magnitude 3.8 **Theta^1 Tauri** to form a wide naked eye double star easily seen just south of brilliant Aldebaran (Alpha Tauri). Aldebaran, a type-K orange giant star, is not a true member of the Hyades, but simply superimposed in front of the cluster. While the Hyades cluster is believed to be about 150 light years away (making it the closest open cluster to Earth after the Ursa Major Moving Cluster), Aldebaran lies at less than half that distance from Earth.

NGC 1647 is found about 3° northeast of Aldebaran. Of the 200 stars that belong in this 1,800-light-year-distant swarm, only 15 are bright enough to be detected readily through 7× binoculars. The remainder meld together into a faint, misty background glow. All of the stars are loosely gathered across a wide 45' sky zone.

NGC 1746, another large open cluster, is bright enough to be within naked eye range on the clearest nights. Twenty stars are identified as true associates of the cluster, with many visible in binoculars. A small clump of faint stars within NGC 1746 has been identified separately in the *New General Catalogue* as NGC 1750, while a portion of the cluster's eastern edge is referred to as NGC 1758. In reality, all three entries pertain to a single cluster.

NGC 1807 and **NGC 1817** lie near the border shared with Orion. The former consists of about 20 stars and has an apparent diameter of 17'. Giant binoculars may just be able to discern the brighter cluster stars set in a cruciform.

NGC 1817 is a less distinct object than its neighbor, even though it has about three times as many stars. The brightest stars found within only reach 10th magnitude and typically are lost in a faint glow when viewed through binoculars.

M1 (NGC 1952), nicknamed the Crab Nebula for its many spiny, nebulous projections visible in photographs, is one of the most fascinating and mysterious objects found anywhere in the heavens. Although M1 was discovered in 1731 by London physician John Bevis, its true nature has only recently come to light. Studies show that the Crab Nebula is all that remains of a very massive star that detonated in a tremendous supernova cataclysm visible from Earth in A.D. 1054.

Buried deep within the Crab is the rapidly beating heart of that ancient star: the Crab Nebula pulsar. The Crab pulsar rotates 30 times each second, sweeping a beam of energy across the Earth with every pass, and remains one of the most swiftly rotating objects of its kind known.

Through 7× binoculars, sharp-eyed observers can spot the faint glimmer of M1 as an 8th-magnitude oval disk of uniform intensity near Zeta Tauri. Larger binoculars add a bit of definition to its amorphous shape, but only hint at the cloud's complex structure.

Dolidze-Dzimselejsvili 4 is a bright open cluster that was apparently missed by Messier, the Herschels, and other great deep sky observers of the past. It is found about 4° east-northeast of the Crab Nebula, where binoculars display about ten luminaries set in a rectangular pattern. In all, Dolidze-Dzimselejsvili 4 hosts about two dozen stars from 6th to less than 10th magnitude.

Telescopium (Tel)

Object	Typ	R.A. (2000) h m	Dec. (2000) ° '	Mag.	Size/Sep/ Period	Comments
NGC 6584	GC	18 18.6	−52 13	9.2	8'	
Hrr 8	OC	18 30.4	−46 08		15'	"X-Marks-the-Spot" Cluster
Delta^{1+2}	**	18 31.8	−45 55	5.0,5.1	9'	Optical
BL	Vr	19 06.6	−51 25	7.7–9.8p	778.1 days	Eclipsing Binary
RX	Vr	19 07.0	−45 58	8.9–10.4p	349.6 days	Semi-Regular
h 5114	**	19 27.8	−54 20	5.7,8.4	70"	248° (1940)
HO	Vr	19 52.0	−46 52	7.9–8.4p	1.613 days	Eclipsing Binary
RR	Vr	20 04.2	−55 43	6.5–16.5p		Z And type;flared to m6.5 in 1944
R	Vr	20 14.7	−46 58	7.6–14.8	461.88 days	Long Period Variable

Introduced by Lacaille in 1752, **Telescopium** occupies a large, indistinct section of the southern summer sky along the eastern shore of the Milky Way. Admittedly, it is a bit difficult to pick out a heavenly telescope among its stars since the three brightest appear to form a right triangle. Perhaps the telescope's tube is bent!

Among the deep sky objects in Telescopium, we find a dim globular cluster, a pair of wide double stars, and several variables.

Delta[1+2] **Telescopii** form an easily seen pair of 5th-magnitude suns in northeastern Telescopium. Separated by 9′ of arc, both shine with the dazzling blue-white radiance of type-B suns. In reality, they are far from each other and are only aligned by chance from our Earth-based vantage point.

Harrington 8 is a curious little asterism of 9th-magnitude stars seen just south of Delta[1]. A more appropriate name might be the X-Marks-the-Spot cluster, as the seven stars in it form the letter "X." Take a look for it through 10×70 or larger glasses.

h 5114 is a second double star in Telescopium that is separable through binoculars. The primary sun, a magnitude 5.7 type-K star, shines with an orange twinkle, while its magnitude 8.4 type-G companion glows a dim yellow. The two may be found about ¾° due east of Eta Telescopii, with the B star seen to the A star's southwest.

Triangulum (Tri)

Object	Typ	R.A. (2000) h m	Dec. (2000) ° ′	Mag.	Size/Sep/ Period	Comments
NGC 598	Gx	01 33.9	+30 39	6.3	60′ × 35′	Sc M33
NGC 604	DN	01 34.5	+30 48			In M33
Cr 21	OC	01 50.1	+27 15	8.2p	6′	Putter Cluster
R	Vr	02 37.0	+34 16	5.4–12.6	266.48 days	Long Period Variable
W	Vr	02 41.5	+34 31	8.5–9.7p	108 days	Semi-Regular

Nestled to the south of Andromeda in our autumn evening sky is the diminutive constellation **Triangulum**. Although it contains no bright stars, the celestial Triangle is easy to pick out thanks to its simple shape and stark surroundings. Within are many galaxies too faint to spot with a pair of common binoculars . . . that is, except for one. One galaxy stands out in this otherwise barren region.

M33 (NGC 598) is that galaxy. Located about one-third of the way from Alpha Trianguli to Beta Andromedae, this is one of the finest

Figure 7.29
The Pinwheel Galaxy M33 (NGC 598) in Triangulum, as photographed by George Viscome through a 3-inch f/6.6 refractor. Exposure time was 21 minutes on ISO 400 film.

examples of a face-on Sc spiral found anywhere in our sky. Photographs (Figure 7.29) display graceful arms stretching out from the galactic core. M33 is famous as a member of the Local Group of galaxies to which our Milky Way belongs. Its distance is projected to be 2.3 million light years.

While its beauty is undeniable, M33 has also earned the reputation of being notoriously difficult to find. Its brightness of magnitude 6.3 can be very misleading since that figure is the galaxy's equivalent brightness if it could be reduced down to a stellar point. However, M33 spreads over almost a full degree, causing the brightness per unit area to be exceedingly low.

In the case of low-surface-brightness objects like M33, wide-field binoculars have a great advantage over narrow-field telescopes. Instead of viewing only a small portion of a large target, binoculars view the entire object in a single field. This ability permits much easier isolation of the object against the sky. I can recall many less-than-perfect nights when M33 was all but invisible through my 8-inch telescope, yet it was clearly seen through 7×50 binoculars.

NGC 604 is the largest of over two dozen diffuse nebulae identified within M33. While M33 is infamous for its low surface brightness, NGC 604 stands out surprisingly well. In fact, under slightly hazy skies, the nebula may actually be visible while the galaxy itself is nowhere to be found! Visible in 7× binoculars as a small, slightly fuzzy "star" 10′ of arc northeast of M33's core, NGC 604 is estimated to be over 1,000 times the size of M42, the Orion Nebula. Powering this mammoth HII region is a cluster of scorching blue-white stars, visible only as specks through the largest telescopes.

Collinder 21 is a pleasant little cluster of 8th- and 9th-magnitude stars found about 2½° southwest of Alpha Trianguli, the Triangle's tip. A friend introduced Collinder 21 to me as the "Putter Cluster" due to its unusual resemblance to a golf putter. Seven-power glasses show only the group's four brightest stars, while 11× and larger glasses continue to fill out the pattern a bit. In all, 20 stars make up the Putter. Anyone want to play through?

Triangulum Australe (TrA)

Object	Typ	R.A. (2000) h m	Dec. (2000) ° ′	Mag.	Size/Sep/ Period	Comments
X	Vr	15 14.3	−70 05	8.1–9.1p		Irregular
R	Vr	15 19.8	−66 30	6.4–6.9	3.389 days	Cepheid
S	Vr	16 01.2	−63 47	6.1–6.8	6.323 days	Cepheid
NGC 6025	OC	16 03.7	−60 30	5.1	12′	
U	Vr	16 07.3	−62 55	7.5–8.3	2.568 days	Cepheid

Far down in the southern summer sky, below Ara and Norma and east of Circinus, is a bright equilateral triangle of 2nd- and 3rd-magnitude stars. This is **Triangulum Australe**, or the Southern Triangle. First introduced in Bayer's *Uranometria* star atlas, the constellation resides almost wholly along the southern boundary of the summer Milky Way. While the area is a fine one for a casual scan through wide-field binoculars, there are surprisingly few deep sky objects that are suitable for binoculars.

S Trianguli Australis is one of three bright Cepheid variables within the Southern Triangle. Located about ½° southeast of Beta Trianguli Australis, it is readily observable across its 6.3-day cycle in even the smallest opera glasses. Glowing with a distinctive deep red hue, this type-M star undulates between a peak of magnitude 6.1 and a minimum of magnitude 6.8.

NGC 6025 is a marvelous splash of stars that bridges the Triangulum Australe–Norma border, but remains affiliated with the former. Binoculars reveal a rich throng of about two dozen 9th-magnitude stars highlighted by a single 7th-magnitude beacon. Encompassing all of them is a small nebulous pool of light created by about 35 fainter cluster stars. Look for NGC 6025 about 1° east of a small triangular asterism of four magnitude 6 suns.

Tucana (Tuc)

Object	Typ	R.A. (2000) h m	Dec. (2000) ° '	Mag.	Size/Sep/ Period	Comments
NGC 104	GC	00 24.1	−72 05	4.5	31'	47 Tucanae
Beta[1+2]	**	00 31.5	−62 58	4.4,4.5	27"	169° (1952)
NGC 292	Gx	00 52.7	−72 50	2.3	280' × 160'	SBMp; Small Magellanic Cloud
NGC 346	OC	00 59.1	−72 11		5'	In SMC
NGC 362	GC	01 03.2	−70 51	6.6	13'	
NGC 371	OC	01 03.3	−72 05		8'	In SMC
T	Vr	22 40.6	−61 33	7.7–13.8	250.76 days	Long Period Variable
Δ 247	**	23 18.0	−61 00	6.7,7.8	47"	291° (1959) Yellow-white

Tucana, the Toucan, is a far southern constellation that is best seen from September through November. First introduced by Bayer in 1603, its placement not far from the south celestial pole never permits the Toucan to fly over much of the northern hemisphere. Still, most readers have undoubtedly heard of some of the deep sky masterpieces that lie within—objects such as the Small Magellanic Cloud and the bright globular cluster 47 Tucanae. For observers situated below the Earth's equator, Tucana is a wondrous place to visit.

NGC 104, better known as 47 Tucanae, lies close to the northwestern edge of the Small Magellanic Cloud. This extraordinary globular cluster is regarded by most who have seen it as perhaps the second finest of its kind, surpassed only by dazzling Omega Centauri. Visible to the unaided eye as a dim, blurry "star," it looks like a large ball of fluff through most binoculars. Eleven-power and larger instruments add a bit of graininess to its surface, implying that resolution of some of its multitude of stars is imminent. The brightest stars in NGC 104 are about magnitude 11.5, which will probably restrict their detection to 4-inch and larger telescopes. The distance to 47 Tucanae is thought to be about 15,000 light years, which is a bit closer than its rival in Centaurus.

NGC 292 (Figure 7.30), the Small Magellanic Cloud or Nubecula Minor, is one of the closest members of the Local Group of galaxies to the Milky Way. It is visible to the naked eye as a hazy cloud about $4° \times 3°$ across and forms an intriguing pair of objects with the Large Magellanic Cloud in Dorado, some 22° to its east. This corresponds to an actual separation of about 80,000 light years.

Some observers using binoculars have likened the curved shape of Nubecula Minor to a fishhook or a comma, with a broader, brighter portion of the Cloud extending toward the south. Through most glasses, we find mostly a featureless surface marked by only four or five faint nebulosities toward the northern edge. Of these, **NGC 346** is the most obvious, lying about 1° northeast of the galaxy's center. Although binoculars fail to identify NGC 346's true nature, we know it to be a cluster of intrinsically very bright stars intertwined with patches of bright nebulosity. Another open cluster visible in giant glasses is **NGC 371,** which lies a bit north and east of NGC 346. Once again, binoculars are not likely to resolve any of the massive giant stars that make up this distant group.

NGC 362 is appreciated by few observers. Were it visible from the northern hemisphere, it would be well known as one of the premier globular clusters. Unfortunately, given its far southern declination as well as the tough competition from both the Small Magellanic Cloud and 47 Tucanae, NGC 362 is destined to remain an unsung showpiece.

Binoculars easily unveil NGC 362 as a 6th-magnitude smudge of light just off the northern edge of the Small Magellanic Cloud. Photographically, it spans 13′ of arc, but to the observer with binoculars, it appears only about half that size. A brighter central core highlights the cluster's fainter surrounding disk.

Figure 7.30
The Small Magellanic Cloud NGC 292 (left), with the dramatic globular clusters 47 Tucanae (NGC 104) lying to its west and NGC 362 to the north. Photograph by *Sky and Telescope's* Dennis diCicco, using an 8-inch f/1.5 Schmidt camera and ISO 400 film.

Ursa Major (UMa)

Object	Typ	R.A. (2000) h m	Dec. (2000) ° '	Mag.	Size/Sep/ Period	Comments
NGC 2841	Gx	09 22.0	+50 58	9.3	8' × 4'	Sb
S 598	**	09 28.7	+45 36	5.5,8,10	77",84"	162° (1924)
W	Vr	09 43.8	+55 57	7.9–8.6	0.334 days	Eclipsing Binary
NGC 2976	Gx	09 47.3	+67 55	10.2	5' × 3'	Scp
NGC 3031	Gx	09 55.6	+69 04	7.0	26' × 14'	Sb M81
NGC 3034	Gx	09 55.8	+69 41	8.4	11' × 5'	P M82
NGC 3184	Gx	10 18.3	+41 25	9.8	7' × 7'	Sc
R	Vr	10 44.6	+68 47	6.7–13.4	301.68 days	M
VY	Vr	10 45.1	+67 25	5.9–6.5		Irregular
TX	Vr	10 45.3	+45 34	7.1–8.8	3.063 days	Eclipsing Binary
VW	Vr	10 59.0	+69 59	6.9–7.7	125 days	Semi-Regular
NGC 3556	Gx	11 11.5	+55 40	10.1	8' × 3'	Sc M108
ST	Vr	11 27.8	+45 11	7.7–9.5	81 days	Semi-Regular
TV	Vr	11 45.6	+35 54	8.3–9.2p	50.4 days	Semi-Regular
65	**	11 55.1	+46 29	6.7,6.5	63"	114° (1969); 8347
Z	Vr	11 56.5	+57 52	7.9–10.8p	196 days	Semi-Regular
Cr 285	OC	12 03	+58	0.4	1,400'	UMa Moving Cluster
RY	Vr	12 20.5	+61 19	6.7–8.5	311 days	Semi-Regular
Wnc 4	**	12 22.4	+58 05	9.0,9.3	50"	M40
T	Vr	12 36.4	+59 29	6.6–13.4	256.24 days	M
S	Vr	12 43.9	+61 06	7.0–12.4	226.02 days	M
Zeta and 80	**	13 23.9	+54 56	2.3,4.0	709"	71° (1966); 8891; Mizar and Alcor
NGC 5457	Gx	14 03.2	+54 21	7.7	27' × 26'	Sc M101 Pinwheel Galaxy
Σ 1831	**	14 16.2	+56 43	7.1,6.6	108"	222° (1956); 9197; Optical

The Big Dipper is without question the northern sky's best known asterism. As we look through the Dipper's home constellation of **Ursa Major,** the Great Bear, our gaze is traveling out beyond the farthest fringes of the Milky Way and into the vast empty void of intergalactic space. Scattered throughout the Bear are hordes of distant galaxies, some quite similar to our own, others amazingly different.

M81 (NGC 3031) and **M82 (NGC 3034)** form one of the most intriguing pairs of galaxies found anywhere in the sky. They may be easily located by extending an imaginary line from Gamma Ursae Majoris (Phecda) to Alpha Ursae Majoris (Dubhe) and then continuing an equal distance to the northwest. There, you will encounter these two striking objects.

M81 is the brighter of the pair by more than a full magnitude. A classic example of an Sb spiral, binoculars as small as 6×30's show its oval form. Increasing to 11×80's heightens the contrast between the galaxy's bright central nucleus and the dimmer surrounding halo of the spiral arms.

M82 is a "peculiar" galaxy, a strikingly different kind of beast. The brightest and most famous of its species, M82 reveals its well-known cigar-like shape through 7× binoculars. Giant glasses even begin to add hints of galactic texture. Using averted vision while peering through 11×80's, I have seen the well-known M82 dark rift that appears to slice the galaxy in half.

Deep photographs reveal a pair of huge nebulous plumes extending from this rift. Many authorities feel that these are clouds of ejecta erupting from within the galactic nucleus. The detection of intense radio noise seems to further advance the theory of an exploding M82, although the studies are not totally compelling.

A third galaxy, **NGC 2976**, is found about 1½° to the south-southwest of M81. At 10th magnitude, 7×50 binoculars will reveal it to patient observers. Although classified as an Sc spiral, NGC 2976 actually shows little spiral structure either visually or photographically.

M108 (**NGC 3556**) was first noticed by Méchain in 1781. Located about a degree south-southwest of Beta Ursae Majoris (Merak), it may be glimpsed through 7× glasses on good nights. Look for an almost pencil-thin smudge of gray light; that will be M108.

While in the neighborhood, can you find planetary nebula M97 (NGC 3587)? Commonly called the Owl Nebula for the two dark "eyes" seen in telescopes, this object really does not belong in a book about the binocular sky. The Owl glows dimly at 11th magnitude but suffers from very low surface brightness due to a large 3' diameter disk. I mention it only because it lies in the same binocular field as M108.

Collinder 285 is the closest and one of the sparsest open clusters visible from Earth. It is without a doubt the most popular cluster in the northern sky. Nearly every resident of the northern hemisphere has seen it at least once, yet few people know of its existence! If this all sounds like a riddle to you, in a way I suppose it is. The five brightest stars in Collinder 285 belong to a much more famous asterism—the Big Dipper itself.

If we could compare the positions of the Dipper's stars 100,000 years from now to how they appear today, we would be hard pressed to identify the familiar figure. Over that stretch of time, even though the familiar bowl and handle pattern will be lost, five of the stars will still move with a common proper motion.

Their shared movement through space was first suspected by R. A. Proctor in 1869 and confirmed three years later by W. Huggins. Studies conclude that at least 16 stars belong to this weak open cluster. The group is about 75 light years away, and spread across an area 18 light years by 30 light years. That translates to an apparent diameter of over 23°. A dozen of the Collinder 285 stars are brighter than 6th magnitude, and are therefore visible without optical aid on clear nights. Although it is impossible to tell at a glance which stars belong to the cluster and which do not, you should be able to pick out the brighter members of the Ursa Major Moving Cluster with the aid of Table 7.3.

M40 (Winnecki 4) has had a checkered past. The trouble began in 1660, when Hevelius recorded a "nebula above the back" of Ursa Major. Try as he might to duplicate the observation a century later, Messier could only find "two stars, very close together and of equal brightness, about 9th magnitude . . . it is presumed that Hevelius mistook these two stars for a nebula." Nevertheless, in 1764, Messier assigned the double star as his catalogue's 40th entry.

In 1863, Hevelius' "nebula" was discovered again, this time by A. Winnecki, who included it as the fourth listing in his double star

TABLE 7.3 Members of the Ursa Major Moving Cluster Brighter than 8th Magnitude

Star	R.A. (2000)			Dec. (2000)			Mag.
21 LMi	10h	07m	25.7s	+35°	14′	41″	4.47
37 UMa	10	35	09.6	+57	04	57	5.16
Beta UMa	11	01	50.4	+56	22	56	2.37
Gamma UMa	11	53	49.7	+53	41	41	2.44
Delta UMa	12	15	25.5	+57	01	57	3.30
HD 111456 UMa	12	48	39.3	+60	19	11	5.87
Epsilon UMa	12	54	01.7	+55	57	35	1.79
78 UMa	13	00	43.7	+56	21	59	4.89
HD 115043 UMa	13	13	36.9	+56	42	28	6.74
Zeta UMa A	13	23	55.5	+54	55	31	2.40
Zeta UMa B	13	23	56.3	+54	55	18	3.96
80 UMa	13	25	13.4	+54	59	17	4.02
Σ 1878 Dra	14	42	03.1	+61	15	43	6.17
Alpha CrB	15	34	41.2	+26	42	53	2.23

inventory. Since that time, Winnecki 4 has shown little change in either separation or position angle (P.A.). The magnitude 9.3 companion star remains 50″ of arc east-northeast of the magnitude 9.0 primary. Under dark skies, both are marginally visible through 7× glasses, while 11× and larger binoculars will have little trouble resolving them. Perhaps you can repeat the original observation of Hevelius, as the stars seem to take on a misty quality when viewed at lower magnifications.

Zeta Ursae Majoris, better known as Mizar, marks the bend in the handle of the Big Dipper. Mizar teams with **Alcor (80 Ursae Majoris)** to form the most famous naked eye optical double star in the entire sky. Separated by nearly 12′ of arc, the pair gleams through even the smallest binoculars. An 8th-magnitude field star also joins Alcor and Mizar, to form a flattened triangular asterism.

While Alcor and Mizar are actually nowhere near each other in space, the latter is one of the most often observed true double stars in the northern sky. Currently, the 4th-magnitude companion may be found about 14″ to the magnitude 2.4 primary's southeast. Since its discovery in 1650, Mizar B has moved little relative to Mizar A. While there is no doubt of the stars' true relationship, the companion's orbit must take thousands of years to complete. Mizar is believed to lie about 74 light years away, corresponding to a minimum separation between the two stars of 318 A.U., or nearly ten times the distance from the Sun to Pluto.

M101 (NGC 5457) is another object that suffers from low surface brightness. As a result, many observers scan right over this face-on Sc spiral without even realizing it. In Messier's own words, M101 is "very obscure and pretty large." Observers stand the best chance of spotting this elusive target if the binoculars are first braced against a steady support. The faint galactic nucleus just might be glimpsed by using averted vision. If you are convinced that your aim is correct, but still nothing is found, try jiggling the glasses slightly. Sometimes marginal diffuse objects are more easily seen in this manner.

Ursa Minor (UMi)

Object	Typ	R.A. (2000) h m	Dec. (2000) ° '	Mag.	Size/Sep/ Period	Comments
Hrr 1	OC	02 32	+89		45'	"Diamond-Ring" asterism
V	Vr	13 38.7	+74 19	8.8–9.9p	72 days	Semi-Regular
U	Vr	14 17.3	+66 48	7.4–12.7	326.51 days	Long Period Variable
Pi1	**	15 29.2	+80 27	6.6,7.3	31"	80° (1959); 9696
S	Vr	15 29.6	+78 38	7.7–12.9	326.25 days	Long Period Variable

Apart from having the Earth's north pole currently aimed almost directly at its brightest star, the constellation **Ursa Minor,** the Little Bear, holds little of interest for amateur astronomers. Its seven primary stars (also collectively known as the Little Dipper) range from 2nd to 5th magnitude, which encourages many amateurs to use Ursa Minor as a gauge to assess the clarity of the night sky. Since Eta Ursae Minoris, the faintest star in the Little Dipper's bowl, shines at magnitude 5.0, its detectability with the naked eye signals a clear sky, at least to us suburbanites.

Harrington 1 is formed from a 45' circlet of stars that includes **Polaris** and nine fainter suns. The renowned *Burnham's Celestial Handbook* points out that, collectively, the stars look almost like a heavenly engagement ring, with Polaris as the diamond. Though not a true cluster, the Diamond-Ring Asterism is worth a look.

S Ursae Minoris and **U Ursae Minoris** are a pair of very similar variable stars. Each fluctuates between approximately magnitude 7.5 to magnitude 13 over a 326-day period. Both of these long period suns cast a reddish glint, especially when spied near their maximum brightness.

Vela (Vel)

Object	Typ	R.A. (2000) h m	Dec. (2000) ° '	Mag.	Size/Sep/ Period	Comments
Gamma	**	08 09.5	−47 20	2,4,8,9	41",62",94"	220°, 151°, 141° (1951)
NGC 2547	OC	08 10.7	−49 16	4.7	20'	
AX	Vr	08 10.8	−47 42	7.9–8.5	2.593 days	Cepheid
AI	Vr	08 14.1	−44 34	6.4–7.1	0.112 days	Delta Scuti type

Vel (continued)

Object	Typ	R.A. (2000) h m	Dec. (2000) ° '	Mag.	Size/Sep/ Period	Comments
h 4069	**	08 14.4	−45 50	6.3,9.0	32″	250° (1959)
Pismis 4	OC	08 34.5	−44 16	5.9	18′	
NGC 2626	DN	08 35.6	−40 40		5′	
RZ	Vr	08 37.0	−44 07	6.4–7.6	20.397 days	Cepheid
Ru 64	OC	08 37.4	−40 06		67′	Weak; difficult to spot
IC 2391	OC	08 40.2	−53 04	2.5	50′	**Omicron Velorum cluster**
IC 2395	OC	08 41.1	−48 12	4.6	8′	
SW	Vr	08 43.6	−47 24	7.4–9.0	23.474 days	Cepheid
Cr 197	OC	08 44.7	−41 22	6.7	17′	
SX	Vr	08 44.9	−46 21	8.0–8.7	9.550 days	Cepheid
NGC 2669	OC	08 44.9	−52 58	6.1	12′	
NGC 2670	OC	08 45.5	−48 47	7.8	9′	
Tr 10	OC	08 47.8	−42 29	4.6	15′	
Δ 73	**	08 56.2	−55 32	7.7,8.2	65″	359° (1938)
CV	Vr	09 00.6	−51 33	6.5–7.3p	6.889 days	Eclipsing Binary
RW	Vr	09 20.3	−49 31	8.9->13.1p	451.7 days	Long Period Variable
WY	Vr	09 22.0	−52 34	8.8–10.2p		Z And type
V	Vr	09 22.3	−55 58	7.2–7.9	4.371 days	Cepheid
GL	Vr	09 26.7	−53 31	7.3–8.0	117 days	Semi-Regular
IC 2488	OC	09 27.6	−56 59	7.4p	15′	
Δ 76	**	09 28.6	−45 30	7.8,7.8	61″	98° (1913)
NGC 2910	OC	09 30.4	−52 54	7.2	5′	
S	Vr	09 33.2	−45 13	7.7–9.5	5.934 days	Eclipsing Binary
NGC 2925	OC	09 33.7	−53 26	8.3p	12′	
Z	Vr	09 52.9	−54 11	7.8–14.8	421.56	Long Period Variable
Cr 213	OC	09 54.7	−50 43		17′	
h 4282	**	10 03.2	−52 03	7.7,8.2	48″	199° (1917)
CM	Vr	10 07.5	−53 16	8.7–11.0p	780 days	Semi-Regular
NGC 3132	PN	10 07.7	−40 26	8.2p	84″×53″	**Eight-Burst Nebula**
NGC 3201	GC	10 17.6	−46 25	6.8	18′	
RY	Vr	10 20.7	−55 19	7.9–8.7	28.127 days	Cepheid
NGC 3228	OC	10 21.8	−51 43	6.0	18′	
h 4329	**	10 31.4	−53 43	4.9,8.4	46″	94° (1938)
h 4330	**	10 32.9	−47 00	5.0,8.9	40″	163° (1960)
NGC 3330	OC	10 38.6	−54 09	7.4	7′	
Δ 95	**	10 39.3	−55 36	4.4,6.6	52″	105° (1938)
SV	Vr	10 44.9	−56 17	7.9–9.0	14.097 days	Cepheid

Vela is the fourth and final component of the now-defunct constellation Argo Navis. Representing the sails of that once-mighty vessel, Vela is composed of several stars that are easily seen with the unaided eye, but are too low in winter's southern sky to be seen well by northern hemisphere residents. From south of the equator, however, they are high in the sky and highlighted by the brilliant Milky Way passing through. Thanks to the presence of our galactic plane, many fine clusters, double and variable stars, and even a rather bright planetary nebula are contained in Vela.

Gamma Velorum, at 2nd magnitude, is the brightest star in the constellation. Binoculars reveal it to be a fine multiple star. The most obvious companion to blue-white Gamma A is a 4th-magnitude white star lying to the primary's southwest. Gamma C and D are both found to the southeast of Gamma A, but shine at only 8th and 9th magnitudes, respectively. Therefore, they are to be seen only by sharp-eyed observers using larger binoculars.

NGC 2547 is located south of Lambda Velorum and stands out magnificently in a very rich Milky Way star field. Over 80 stars belong to this dazzling starry array, with more than a dozen shining brighter than 9th magnitude. This group always reminds me of a curved cross lying on its side. If you have never seen NGC 2547 and have the opportunity, do not let it pass you by!

Pismis 4 is a weak, though apparent, open cluster in western Vela. Only a few of its 45 stars are bright enough to be seen through binoculars. The rest fall short of detection, but may produce a weak background glow.

Surrounding Pismis 4 (though not physically associated with it) is the **Gum Nebula.** Named for its discoverer, Colin Gum, the nebula is not commonly thought of as a visual object, as it usually turns up only on long-exposure photographs. Certainly, it cannot be seen in a single telescope field, since it extends over 35 square degrees. With the advent of contrast-enhancing nebula filters, however, it just might be possible to see the Gum Nebula through binoculars. Any takers to the challenge?

IC 2391 can be seen with the unaided eye on good nights as a hazy spot involving many faint stars, including 4th-magnitude Omicron Velorum. The cluster may easily be found about 1½° northwest of Gamma Velorum. Binoculars reveal a large, lustrous gathering containing two dozen stars down to 9th magnitude, including several doubles. Unfortunately, the beauty of IC 2391 remains unknown to most observers, as it never rises above the southern horizon from much of the northern hemisphere. Those amateurs fortunate enough to be located in the south may partake in a memorable treat.

IC 2395 is another bright open cluster doomed to anonymity because of its far southern location in the sky. Consisting of about 40 magnitude 5 and fainter suns, it appears as a small, tight swarm of stardust set to the south of a field abundant with stars. In all, I have counted about a dozen points of light here in 11× binoculars while observing at latitude 25° N. Perhaps southern hemisphere residents will be able to see even more with the cluster higher in the sky.

NGC 2669 is found about a degree due east of Omicron Velorum and IC 2391. Through binoculars, it is a moderately bright glow surrounding six stars set in a trapezoidal pattern. By itself, NGC 2669 would be a noteworthy object, but when combined with larger IC 2391 to its east, as well as its general surroundings, the overall effect is truly magnificent. Unfortunately, once again, neither of the two clusters is ever seen well above latitude 25° N.

Trumpler 10 is a tightly packed swarm of stars more ideally placed than the two previous clusters for observation by northern hemisphere astronomers. Found just east of the star d Velorum, Trumpler 10 claims 40 stars crammed into a crowded 15′-of-arc area. Approximately nine of these are bright enough to be seen in binoculars, including a readily visible 6th-magnitude double star.

IC 2488 is located 2° south of Kappa Velorum and ½° west of N Velorum, riding the Vela-Carina boundary. Over 70 stars belong to this 15′-diameter open cluster, but most shine at fainter than 9th magnitude. Only a few are bright enough to be seen through most glasses, while the rest coalesce into a dim stellar fog totaling 7th magnitude.

NGC 3132 is one of the brightest planetary nebulae in the entire sky. Located exactly on the Vela-Antlia border, it is officially assigned to the former. NGC 3132 shines conspicuously at 8th magnitude and is comparable in size to the famous Ring Nebula in Lyra. Its central star is 10th magnitude and may possibly be seen in 10× glasses. Nicknamed the Eight-Burst Nebula due to its complex multi-ring appearance in photographs, NGC 3132 reveals a whitish disk to visual observers. About 2,600 light years separate it from Earth. The Eight-Burst Nebula may be found by scanning 10° due east of Psi Velorum. To aid in locating it, there is an unrelated 9th-magnitude star just south of the nebula.

NGC 3201 is the only globular cluster found within Vela. It may be found about 5½° northwest of 3rd-magnitude Mu Velorum. Globulars are typically classified on a 12-point scale according to their stellar concentration, with higher values signifying a lower density. On the basis of this scale, NGC 3201 is listed as Class X, implying a loose structure. Binoculars will readily show its disk, but the cluster's thousands of individual stars remain invisible at magnitude 13 and below.

NGC 3228 is a bright galactic cluster in eastern Vela. Through 7× glasses, about a dozen 9th-magnitude stars compressed into an 18′ zone are visible here, with about half crushed into a tight centralized knot. Collectively, the cluster has an integrated magnitude of 6.0, permitting easy detection through binoculars.

Virgo (Vir)

Object	Typ	R.A. (2000) h m	Dec. (2000) ° ′	Mag.	Size/Sep/ Period	Comments
X	Vr	12 01.9	+09 04	7.3–11.2		Unknown type
RW	Vr	12 07.2	−06 46	8.6–9.1p		Irregular
NGC 4216	Gx	12 15.9	+13 09	10.0	8′ × 2′	Sb
NGC 4303	Gx	12 21.9	+04 28	9.7	6′ × 5′	Sc M61
NGC 4374	Gx	12 25.1	+12 53	9.3	5′ × 4′	E1 M84
SS	Vr	12 25.3	+00 48	6.0–9.6	354.66 days	Long Period Variable
NGC 4406	Gx	12 26.2	+12 57	9.2	7′ × 6′	E3 M86
NGC 4472	Gx	12 29.8	+08 00	8.4	9′ × 7′	E4 M49
BK	Vr	12 30.4	+04 25	7.3–8.8	150 days	Semi-Regular
NGC 4486	Gx	12 30.8	+12 24	8.6	7′	E1 M87
NGC 4526	Gx	12 34.0	+07 42	9.6	7′ × 2′	E7
NGC 4535	Gx	12 34.3	+08 12	9.8	7′ × 5′	SBc
NGC 4536	Gx	12 34.5	+02 11	10.4	7′ × 3′	Sc
NGC 4552	Gx	12 35.7	+12 33	9.8	4′	E0 M89
NGC 4569	Gx	12 36.8	+13 10	9.5	9′ × 5′	Sb+ M90
NGC 4579	Gx	12 37.7	+11 49	9.8	5′ × 4′	Sb M58
R	Vr	12 38.5	+06 59	6.0–12.1	145.64 days	Long Period Variable
NGC 4594	Gx	12 40.0	−11 37	8.3	9′ × 4′	Sb M104 Sombrero Galaxy
NGC 4621	Gx	12 42.0	+11 39	9.8	5′ × 3′	E3 M59
NGC 4649	Gx	12 43.7	+11 33	8.8	7′ × 6′	E1 M60
NGC 4697	Gx	12 48.6	−05 48	9.3	6′ × 4′	E4
NGC 4699	Gx	12 49.0	−08 40	9.6	4′ × 3′	Sa
U	Vr	12 51.1	+05 33	7.5–13.5	206.80 days	Long Period Variable
NGC 4753	Gx	12 52.4	−01 12	9.9	5′ × 3′	P
RT	Vr	13 02.6	+05 11	9.0–10.3p	155 days	Semi-Regular
SW	Vr	13 14.1	−02 48	6.9–7.9	150 days	Semi-Regular

Vir (continued)

Object	Typ	R.A. (2000) h　m	Dec. (2000) °　'	Mag.	Size/Sep/ Period	Comments
S	Vr	13 33.0	−07 12	6.3–13.2	377.43 days	Long Period Variable
DL	Vr	13 52.6	−18 43	7.0–7.5	1.315 days	Eclipsing Binary
RS	Vr	14 27.3	+04 41	7.0–14.4	352.80 days	Long Period Variable

Welcome to galaxy country! **Virgo**, the Virgin, is home to more bright galaxies than any other constellation in the sky. Most belong to a colossal galactic swarm called the *Virgo Realm of Galaxies*.

The Virgo Realm of Galaxies is the closest and one of the largest assemblages of galaxies in the known universe. Studies indicate that at least 2,500 individual galaxies are found within the Realm at an average distance of 78 million light years from the Milky Way. Spanning over 100 square degrees in Virgo and neighboring Coma Berenices, the Realm's actual diameter is about 16 million light years.

The 17 brightest island universes in Virgo's Realm await the roving eye of binocular astronomers. These galaxies are rather faint by binocular standards; they certainly will not poke you in the eye and say "here I am!" Through most glasses, each will challenge the skill of even the most accomplished observer. Still, with patience, their tiny dim glimmers of light can be seen floating amongst the stars.

M61 (NGC 4303) is found 5° north of 4th-magnitude Eta Virginis, although it is difficult to spot in anything less than 10× binoculars. Photographs reveal this 10th-magnitude galaxy to be a face-on spiral measuring about 6′ across; visually, it appears about half that size.

M84 (NGC 4374) teams with **M86 (NGC 4406)** to form a pair of close-set galaxies near the Virgo–Coma Berenices border. Through binoculars, both of these elliptical galaxies appear as nebulous, oval disks punctuated by brighter central nuclei. M86, the westernmost of the two, appears somewhat brighter than M84. They were both discovered by Messier in March 1781.

M49 (NGC 4472) is located a little to the east of the halfway point between M61 and M84/86. Messier saw M49 "only with difficulty" when he happened upon it in February 1771. It leaves me with a somewhat brighter impression, as I feel it stands out fairly well in 7×50's on dark nights. Of course, it is quite a different matter to discover something for the first time than it is to see something you *know* exists. Classified as an E4 elliptical, M49's 8th-magnitude, 9′×7′ oval disk makes it the second brightest galaxy in Virgo.

M87 (NGC 4486) is famous as one of the largest and most luminous galaxies in the entire universe, with estimates indicating a stellar population of approximately 800 billion solar masses. In 1918, Heber D. Curtis of Lick Observatory discovered a luminous jet of material bursting from M87's core. To add to the mystery, strong radio emissions were detected from M87 in 1948, while X-rays from this most unusual object were identified 18 years hence. Today, all three of these unusual features are tied into a single theory stating that the jet results from a massive gravitational collapse from deep within the galactic core, perhaps caused by a black hole.

Binoculars reveal M87 to have a perfectly stellar nucleus engulfed in a fainter round mist. To my eye, it outshines M84 and M86 by about a full magnitude, but most lists claim it is only half a magnitude brighter.

M89 (NGC 4552) and M90 (NGC 4569) form a challenging galactic duo hugging the Virgo-Coma Berenices border. The former, a 10th-magnitude E0 elliptical, can prove difficult to detect through binoculars of less than 10×, especially from a less-than-perfect observing site. Even then, it appears as little more than a minute, featureless smudge without the slightest trace of a central concentration.

M90, found ¾° northeast of M89, is about half a magnitude brighter and somewhat more impressive. With a bit of study, especially through giant glasses, this Sb spiral shows a faint stellar center girdled by the oval nebulosity of its tightly wound arms.

M58 (NGC 4579) lies about a degree southeast of M89. Although they are listed as the same magnitude, I always seem to be able to locate M58 more easily than its neighbor to the north. This is probably due to the slightly smaller sized disk of M58, which concentrates the same amount of light over a lesser area to yield a brighter effect. Once again, observers peering through large binoculars just discern a stellar nucleus buried within a dull, milky haze.

Incidentally, about half a degree southwest of M58 is an interesting pair of interacting galaxies labeled NGC 4567 and NGC 4568 and nicknamed the "Siamese Twins." Although they just break the 11th-magnitude barrier (and are therefore too faint to be included in our survey), they are still of special interest since they form one of the sky's brightest pairs of physically connected galaxies. Most amateur telescopes reveal them as an unresolvable single object with a rather strange shape.

M104 (NGC 4594), the brightest galaxy in Virgo, is found far to the south of the Realm's center. This is the celebrated Sombrero Galaxy, named for its protruding central core and broad, flattened spiral arm rim. Its appearance is further highlighted by a prominent dark lane encircling the galaxy along the arms' outer circumference.

M104 is a marvel to behold in nearly all binoculars. Lower power glasses reveal an oval disk, which brightens rapidly toward the center. With higher magnification comes an even more spectacular view. Through 11×80's, the galaxy's bisecting band of dark nebulosity may be suspected as it slices across the Sombrero's "brim."

M59 (NGC 4621) and **M60 (NGC 4649)** may be found by continuing another two degrees to the southeast from M58. Although both are evident when conditions are right, M59 will pose quite a challenge through less-than-optimum skies. Shining at 10th magnitude, its 5'×3' oval disk will frequently require averted vision to be spied.

M60 appears a full magnitude brighter and a bit larger than M59, and looks just like a tiny ball of fluff. Its nucleus is also suspectible when viewed with averted vision. Both M59 and M60 were discovered in April 1779 by J. Koehler.

Volans (Vol)

Object	Typ	R.A. (2000) h m	Dec. (2000) ° '	Mag.	Size/Sep/ Period	Comments
S	Vr	07 29.8	−73 23	7.7–13.9	395.83 days	Long Period Variable
Kappa¹⁺²	**	08 19.8	−71 31	5.4,5.7	65″	57° (1835)

Volans, the Flying Fish, is another of the many southern hemisphere constellations conceived by the creative mind of Bayer in the early 17th century. Occupying the space bordered by Dorado and the Large Magellanic Cloud to the west and Carina to the east, Volans holds but two deep sky objects that are bright enough for binocular scrutiny. Many galaxies also dwell in the area, but they are undetectable through binoculars.

Kappa¹⁺² Volantis is a bright, attractive binary star found close to the constellation's center. Kappa¹, a magnitude 5.4 type-B star, is isolated from magnitude 5.7 Kappa², a type-A sun, by more than a full minute of arc. This wide tract of sky between them makes the pair a picturesque binary system when viewed with almost any optical aid. Both Kappa¹ and Kappa² glisten pure white against a dark sky far from any stellar competition.

Vulpecula (Vul)

Object	Typ	R.A. (2000) h m	Dec. (2000) ° '	Mag.	Size/Sep/ Period	Comments
RS	Vr	19 17.7	+22 26	6.9–7.6p	4.478 days	Eclipsing Binary
Z	Vr	19 21.7	+25 34	7.4–9.2p	2.455 days	Eclipsing Binary
Cr 399	OC	19 25.4	+20 11	3.6	60'	**Coathanger Cluster**
NGC 6800	OC	19 27.2	+25 08			
Stock 1	OC	19 35.8	+25 13	5.3	60'	
U	Vr	19 36.6	+20 20	6.8–7.5	7.991 days	Cepheid
NGC 6815	OC	19 40.9	+26 51			
NGC 6820	DN	19 43.1	+23 17		40'	
NGC 6823	OC	19 43.1	+23 18	7.1	12'	
NGC 6830	OC	19 51.0	+23 04	7.9	12'	
SV	Vr	19 51.5	+27 28	6.7–7.8	45.035 days	Cepheid
NGC 6853	PN	19 59.6	+22 43	8.1	480"×240"	**M27 Dumbbell Nebula**
NGC 6885	OC	20 12.0	+26 29	5.7p	7'	
NGC 6940	OC	20 34.6	+28 18	6.3	31'	
FI	Vr	20 48.9	+23 00	8.6–9.4p		Irregular
T	Vr	20 51.5	+28 15	5.4–6.1	4.436 days	Cepheid
R	Vr	21 04.4	+23 49	7.0–14.3	136.36 days	Long Period Variable

Vulpecula, the Fox, is a small summertime constellation found drenched in the Milky Way, yet it offers no stars brighter than magnitude 4.5. For an observer armed with binoculars, however, this region blossoms forth with many deep sky splendors.

Collinder 399, seen above the center in Figure 7.31, is a fine cluster of stars when viewed through binoculars, yet many observers seem to overlook it. You may know it by one of its more popular nicknames, the Coathanger Cluster or Brocchi's Cluster. The Coathanger contains some five dozen stars spanning a full degree of sky, and may be seen without any optical aid as a bright patch within the Milky Way. The cluster's brightest suns, labelled separately as 4, 5, and 7 Vulpeculae, are also visible to the unaided eye if conditions permit.

Binoculars easily reveal how this cluster came to be known as the heavenly Coathanger. Six stars, aligned in a straight line, form the coathanger's cross bar, while four other points of light curve away to create its hook. Seven-power binoculars display a total of 20 stars sprinkled across Collinder 399. Surrounded by a magnificent field filled with stardust, this is sure to become one of your favorite summertime objects, just as it has become one of mine.

Figure 7.31
The Coathanger Cluster, Collinder 399, in Vulpecula appears just above the center of this photograph, while Altair (in neighboring Aquila) is the bright star below left center. The author took this ten-minute exposure using a 50-mm f/1.7 lens on ISO 400 film.

Stock 1 is a second large galactic cluster inside of Vulpecula. Studies indicate that 40 stars may be found within its 1°-diameter boundaries. About half of these are 9th magnitude or brighter, and are therefore visible in 7×50 glasses. However, unlike Collinder 399, it is difficult to tell exactly where Stock 1 begins and ends. It impresses me more as a rich star field than a true cluster.

To find Stock 1, scan about 3½° southeast of Albireo (Beta Cygni) for a distinctive, though faint, arc of four stars positioned just west of an 8th-magnitude orb. These suns are centrally located within the cluster. A 7th-magnitude star to the arc's northwest stands out as the group's brightest luminary.

NGC 6823, an open cluster found about 6° southeast of Albireo (Beta Cygni), consists of 30 magnitude 9 and fainter stars. Most binoculars reveal only four separate points of light amid a soft glow of other unresolved cluster members.

Under ideal conditions, it might also be possible to detect **NGC 6820,** a large diffuse nebula. Extending for about half a degree around the cluster, this cloud's low surface brightness makes it a most difficult object to run down.

M27 (NGC 6853), the Dumbbell Nebula (Figure 7.32), is famous as one of the finest planetary nebulae around. Many observers, including myself, feel this is the most spectacular of its kind in the entire sky. Messier was first to happen upon the Dumbbell Nebula in July 1764. Although he could not imagine its origin, he correctly observed that it was nonstellar in appearance. Its nickname comes from its similarity in appearance to a closely set pair of weights joined by a short bar.

Equal to an 8th-magnitude star in brightness, the Dumbbell may be seen through 7× glasses as a hazy, rectangular smudge just south of 14 Vulpeculae. Higher magnification begins to reveal its celebrated hourglass form and azure color.

Astronomers estimate that the Dumbbell Nebula lies about 975 light years away, making it one of the nearer planetaries. This correlates to an actual size for M27 of about 2½ light years across. Given these facts, and a measured expansion rate of approximately 17 miles per second, we can work backwards to determine the age of M27, which works out to be about 48,000 years. This is more than twice the age of typical planetaries.

NGC 6940 is a striking open cluster lying just inside the northern border of Vulpecula. Here, observers will find about five or six feeble

Figure 7.32
The Dumbbell Nebula, M27 (NGC 6853), in Vulpecula. Photograph by
Lee Coombs through a 610-mm f/6 lens using 103a-E spectroscopic film
and a ten-minute exposure.

pinpoints against the combined glow of the cluster's five dozen suns.
One of these faint stars is known as FG Vulpeculae, a red semi-regular
variable star. It ranges in brightness from magnitude 9.0 to magnitude
9.5 across approximately 80 days.

Appendix A

Caveat Emptor! (Let the Buyer Beware!)

Anyone can buy a pair of binoculars, but buying the right pair for your needs requires planning and forethought. The binoculars you choose will become an extension of your senses, so you want the partnership to be a long and happy one. Before you begin to shop, consider a few factors.

Modern binoculars are available in two basic styles, depending on the type of internal prisms used: porro prism glasses and roof prism glasses (Figure A.1). Roof prism glasses, although more compact in design than porro prism glasses, do not lend themselves as well to astronomical applications. The problem is that roof prism assemblies do not reflect light as efficiently as porro prisms. By design, roof prisms require one internal surface to be aluminized in order to bounce light through the binoculars. As the light strikes the aluminized surface, a bit of it is lost, resulting in a dimmed image. Porro prisms—at least those made from high-quality glass—totally reflect the incoming light without the need for a mirrored surface, allowing for a brighter final image.

The second disadvantage to roof prism glasses is their high cost. Part of the higher expense is the result of the precision

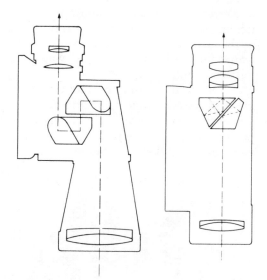

Figure A.1
Cutaway diagram showing internal components of porro prism (left) versus roof prism (right) binoculars. Courtesy of Unitron Corporation.

required for their assembly: unless constructed with great care, the image quality of roof prism binoculars will be unacceptable. Still, for viewing the night sky, even the

263

best roof prism glasses are inferior to porro prism models.

Porro prism binoculars come in two body styles: German and American (Figure A.2). The German style consists of a two-piece body, with the objective lenses held in separate barrels from the barrels holding the prisms and eyepieces. Due to the joint between the two pieces, German glasses may be knocked out of alignment with only a slight jolt. In addition, the construction joints provide poor seals against the infiltration of dust and moisture. It is not unusual for German-style glasses to suffer from internal fogging, especially during times of high humidity.

American-style binoculars consist of a one-piece body. Their design offers a sturdier method of holding the optics and maintaining alignment. Lacking the central joint of the German-style glasses, American binoculars more thoroughly seal out water and dust contamination.

One of the most important choices to make in selecting a new pair of binoculars is the magnifying power of the glasses. Low-power binoculars are great for wide-field views of the Milky Way, but higher power glasses are preferred for the Moon, Sun, and planets. Keep in mind that as magnification increases, so does the binoculars' weight. This is especially important if you plan on holding the glasses by hand. Most people can support 7× binoculars with minimal strain and fatigue, but with 10× glasses, the extra weight might make hand support impractical.

Binoculars over 10× are classified as *giant binoculars*. These are especially useful in comet hunting, lunar and planetary examination, and any application that requires a little more power than traditional glasses.

If you have your heart set on purchasing a high-power pair, then some sort of external support (typically a photographic tripod) is a must. Appendix B evaluates methods for mounting binoculars and offers a sampling of alternatives to traditional tripods constructed by some amateur astronomers.

Another critical choice is the binoculars' objective lens diameter. When it comes to selecting a telescope, amateur astronomers always assume that "bigger is better." While this is generally true for telescopes, it

Figure A.2
American-style (one-piece) versus German-style (two-piece) binoculars. Courtesy of Unitron Corporation.

is not always the case for binoculars. For optimum performance, the objective diameters should be matched to both your eye and your observing site.

The human eye is an amazing optical mechanism. It can automatically adapt itself to a wide range of varying light conditions. The amount of light entering the eye is controlled by the central opening in the iris called the *pupil*. For most people, the pupil contracts to about 2½-mm in broad daylight and dilates to about 7-mm in total darkness. To be perfectly matched for nighttime sky watching, the cone of light leaving the binoculars' eyepieces, known as the *exit pupil*, should be the same diameter as the diameter of the eye's pupil. If the exit pupil is too small, then some of the eye's light-gathering capability is wasted; with too large an exit pupil, we are physically unable to make use of the binoculars' full potential.

The exit pupil for a particular pair of binoculars is determined by dividing the objective's diameter by the power of the binoculars. For example, a 7×50 pair has a 7.1-mm exit pupil, while 7×35 glasses have a 5-mm exit pupil.

The binoculars' objective diameter and resulting exit pupil should be closely related to the conditions under which the binoculars will most frequently be used. If you spend most of your observing time under dark, rural skies, then your eyes' pupils should dilate fully to about 7-mm. Therefore, a diameter-power combination producing a matching 7-mm exit pupil should be chosen. Among these are 7×50, 10×70, and 11×80 binoculars.

If your location suffers from light pollution, then your pupils will never become fully dilated. Instead, they may only expand to 5-mm or 6-mm, making night glasses with large exit pupils unnecessary. In fact, along with increased aperture may come increased sky glow interference. The larger diameter objectives will pick up more light pollution along with more starlight and may

actually produce images inferior to those produced by smaller binoculars. Accordingly, urban and suburban residents might do better by selecting binoculars with smaller exit pupils, such as 7×35's or 10×50's.

Most binocular manufacturers offer zoom binoculars. These permit the user instantly to double or even triple the power of the instrument. Unfortunately, nothing in this life is free, and zoom binoculars pay heavily. To perform this feat, zoom models require far more complex optical systems than fixed-power glasses. More optical elements increase the risk of imperfections and inferior performance. Additionally, image brightness suffers terribly at the high end of zoom binoculars' magnification range, causing faint objects to vanish.

In purchasing binoculars, it is also important to know the binoculars' field of view. A value for the field of view is usually found stamped on the barrels' tail stock and may be expressed in two different ways. Typically, manufacturers will either note the area of view in degrees or express it as a separation in feet at 1,000 yards (e.g., 325 feet at 1,000 yards). The latter tells us that if a 325-foot-long ruler were viewed through the glasses from a thousand yards away, it would just squeeze into the eyepieces' field of view. If your binoculars specify the field of view by this second method, you may determine their true angular field by dividing the number of feet by 52.5. Thus, the view through the "325-feet" binoculars actually covers a little over 6° of sky.

Wide-angle binoculars are becoming increasingly popular. These binoculars may give up to a 10° to 12° field, permitting easy detection of objects over a broad area. Wide-angle glasses are available in the same magnification-diameter combinations as normal-angle glasses, but require larger and heavier prisms and eyepieces. Accordingly, they are more difficult to make and more expensive to purchase. So-called extra-wide-angle binoculars should be looked upon with a suspicious eye. Glasses that offer fields greater than about 600 feet at 1,000 yards (11.5°) often suffer badly from coma (making stars near the edge of the field look like little comets) and other optical imperfections. Of course, all optical systems exhibit some coma, but it generally becomes more pronounced as the angular field of view expands.

Table A.1 summarizes some of the more popular binocular sizes and their capabilities.

When shopping for a pair of binoculars, it is a good idea to compare several brands side by side. Look at the objective lenses. Does the manufacturer state that the lenses are coated? Optical coatings reduce lens flare and improve light transmission, two desirable characteristics for astronomical binoculars.

A plain, uncoated lens actually reflects about 4% of the light striking it. By applying an antireflection coating to both sides of the lens, light reflection is reduced to about 1½%. Most optical instruments available to-

TABLE A.1 Comparison of Binoculars

Size	Advantages	Disadvantages
6 × 30	Small size, light weight.	Power too low to reveal many celestial objects.
7 × 35	Easily hand-held; excellent wide-field views of Milky Way and deep sky objects.	Exit pupil restricts dark-sky effectiveness.
7 × 50	Easily hand-held; light-gathering ability sufficient for hundreds of objects; best overall choice.	Increased objective diameter may cause sky-glow problems in urban and suburban areas.
10 × 50	Good choice for urban or suburban users who want a little higher magnification; added power reveals many details on the Moon.	May require a tripod.
10 × 70	Excellent for star clusters, nebulae, and galaxies.	Heavy; tripod usually needed.
11 × 80	Excellent for faint objects; stereoscopic views of interstellar regions beyond comparison.	Tripod needed; heavy; can be clumsy to carry and use.
15 × 80 ⎫ 20 × 70 ⎬ 20 × 80 ⎭	High power good for lunar and planetary viewing.	Restricted exit pupil might be a problem for dark-sky viewing.
30 × 80	Limited to bright objects, such as planets and the Moon.	Not intended for astronomical use due to small exit pupil.

day, such as camera lenses and binoculars, are coated with magnesium fluoride, which is applied by evaporation to the glass surface in a vacuum chamber.

The images produced by binoculars with uncoated lenses appear far inferior to those produced by their coated counterparts. The runaway reflected light strays throughout the optical system, causing low-contrast images against a muddy background.

Some moderately priced binoculars that state that they have "coated optics" should be interpreted to mean, "Only the exposed faces of the objectives and eye lenses are coated"; the internal optics are probably not. Others may claim that all air-to-glass optics are coated, which implies that all internal surfaces are coated as well. However, the manufacturer may "forget" to inform the prospective consumer that some of the internal lenses are made of plastic. Chances are that those nonglass elements are uncoated.

The thickness of the applied coating is critical to its effectiveness. For magnesium fluoride, the ideal thickness is ¼ of the wavelength of the light passing through. The human eye is sensitive to all light across the visible spectrum ranging from red to blue, but especially to the color in the middle: yellow. Yellow has a wavelength of 550 nanometers (one nanometer equals 10^{-9} meter). Manufacturers aim at 25% of 550 nm when specifying the coating's thickness.

A lens with the proper thickness of magnesium fluoride exhibits a purplish hue when held at a narrow angle toward a light. If the coating is too thin, it will appear pinkish; if it is too thick, then a greenish tint will be cast. Uncoated lenses have a whitish glint. Reject any lenses that are not fully coated with the correct thickness.

Top-of-the-line binoculars receive multiple antireflection coatings, reducing reflection to less than ½%. Most multicoated lenses show a greenish reflection when

turned toward a light. However, some less-than-honest dealers may try to pass off a lens with too thick a single coating as multicoated binoculars, as both can appear very similar.

When shopping, carefully check the binoculars' optical alignment. The brain is used to getting two nearly identical images from our eyes. Any misalignment between the binoculars will result in two slightly off-axis images. In an effort to correct this apparent malfunction, the brain will strain to bring the images together. As a result, misaligned binoculars will cause undue eye strain and headaches when used for any significant length of time. In his book, *Binoculars and All-Purpose Telescopes* (New York: AMPHOTO, 1980), the late Dr. Henry Paul recommends a simple way to check the alignment of binoculars quickly. While looking at a distant terrestrial object, cover one side of the binoculars with your palm, but keep both eyes open. Then quickly remove your hand to uncover the lens. If any misalignment exists, two images will be seen for an instant before the brain tries to snap the images together.

Now hold the binoculars at arm's length, and look at the circle of light coming out of each eyepiece (Figure A.3). Are they both clear circles, or do you notice a diamond shape within? Clear circles indicate that the pair's prisms are made of BAK-4, a superior-quality glass. The index of refraction of this type of glass allows for total internal reflection of all light entering the prisms. The diamond effect is caused by light falloff from using cheaper Crown glass (BK-7) prisms. Crown glass does not permit total internal reflection, which forces manufacturers to aluminize one face of the prisms in order to reflect light. The result is dimmer images.

Besides noting the optical quality of the binoculars you are considering purchasing, take a long, hard look at the mechanical assembly, starting with the focusing mecha-

Figure A.3
A comparison between the clear circular exit pupil of binoculars with BAK-4 prisms versus the diamond-like exit pupil from glasses constructed with less expensive BK-7 prisms. Photographs courtesy of Unitron Corporation.

nism. Most contemporary binoculars have a knurled focusing wheel between the barrels, while older binoculars frequently require each eyepiece to be focused separately. Regardless of the method used, the focusing device should move smoothly. Incidentally, some newer binoculars use a "fast-focus" thumb lever instead of a wheel. This approach permits rapid, coarse focusing, which may be terrific for moving objects, but is not accurate enough for the fine focusing required when stargazing.

Next, hold the binoculars up to a light and look into the front end of the barrels through the objectives. Are the lenses free of scratches? The barrel walls should be free of any foreign matter such as dust, adhesive, and metal shavings. If not, the binoculars are unacceptable.

Another item to check is the eyepieces' interpupillary distance adjustment. Binoculars must be adjusted to fit a wide range of observers. To achieve this versatility, glasses have a central hinge that permits the distance between the eyepieces to be adjusted comfortably for your eyes. Also, check whether the particular model you are interested in has a built-in standard photographic tripod socket. In most cases, the front end of the barrel hinge is threaded for this purpose, which makes a right-angle adaptor necessary for mounting the glasses onto a tripod. Rarely, a pair is found with the tripod socket built right into the bottom of one of the barrels. This type of tripod socket should be avoided, as alignment problems may be introduced.

As with nearly everything we buy, the budget will be the final dictating factor. Perhaps you cannot afford the best binoculars around, but after performing these simple tests, you will be able to select the best binoculars within your price range.

Appendix B
Mounting Concerns

One of the basic advantages binoculars have over telescopes is portability: while telescopes may require a half hour or more to set up, it is a simple task to pull binoculars out of their case and begin scanning the heavens.

With the ever-increasing popularity of so-called giant glasses, more and more amateurs are realizing the difficulty in holding binoculars steady for extended periods of time. Whereas most of us have little trouble supporting 7× glasses for short periods while viewing level, terrestrial objects such as sailboats on the horizon, we quickly grow fatigued while searching for a faint celestial object near the zenith. Moreover, as magnification climbs, so will a pair's weight and size, making larger glasses even more difficult to support by hand. Great care may be given to their selection, but without solid support, the performance of even the finest binoculars will be less than satisfactory.

If you want to hold binoculars steady by hand, try this trick. Most people naturally hold binoculars around the prism end of the barrel, which places the entire weight of the objective lenses out in front of the support points (your hands). However, if the binoculars are grasped at the objective (front) end of the barrels, the overhanging weight will be directly supported, thereby greatly reducing shaking. I have been able to steadily hold 20× and even 30× glasses in this manner without the use of a tripod. Extended use will still cause fatigue, however, making some sort of external support necessary.

Not all camera tripods are suitable for supporting binoculars; indeed, most are not suitable for the task. The flimsy legs and weak heads of less expensive tripods prove little better than holding binoculars by hand, especially when used with heavier giant models. A good tripod will have cross-braced telescoping legs that lock firmly in place. Since most of your viewing time will be spent looking up from *under* the binoculars, the tripod's maximum height is also an important consideration. Nothing takes the joy out of stargazing quicker than having to bend and strain just to look through the eyepieces. Depending on your height, the tripod should raise the binoculars six feet or more.

Some of the best binocular mounts are not for sale at any price. Many clever amateur telescope makers have come up with ingenious designs for making observing with binoculars more comfortable and convenient. For example, Jerry Burns, a retired machinist from Moorestown, New Jersey, has designed and constructed the clever, but simple, binocular mounting system shown in Figure B.1. His idea of a "flexible parallelogram" is strongly akin to swing-arm desk lamps. These lamps may be raised and lowered to any position, with the light itself remaining aimed at a constant angle.

Figure B.1
Jerry Burns (right) and his homemade parallel-arm binocular mount and matching tripod. Photograph by the author.

The "flexible parallelogram" idea is one of the nicer features of Burns's system. Once aimed, the binoculars may be raised and lowered to permit comfortable viewing by observers of different heights without ever losing the target. To see how this is possible, notice that, like swing-arm desk lamps, the frame of Burns's construction forms a hinged parallelogram. Thus, even though the angle between adjacent sides may change, the sides opposite each other remain parallel. Further, the shorter vertical sides always hold the same angle relative to the tripod on which they are mounted. This means that anything attached to one of the shorter legs of the parallelogram will stay aimed in one direction even when the frame is raised up or down.

All of the mount's components were machined from scrap aluminum in Burns's home workshop. Each joint is pinned to ensure the close tolerances required for the mount to work properly. Nylon washers are placed between each pinned metal-to-metal pivot to prevent possible binding.

An adjustable counterweight, opposite the binoculars, helps stabilize the mounting. Burns found that, even when the mount was perfectly balanced, some additional frictional drag was required. To supply this drag, he introduced a pair of nylon-tipped threaded rods into the system that pass through two side members and press against an aluminum semicircle attached to the top beam. Pressure may be adjusted by turning one or both of the white sheathed rods by hand.

While his design can be attached to any heavy-duty photographic tripod, Burns chose to build his own tripod from scratch. The resulting combination of homemade binocular mount and tripod is elegant for its simplicity, functionality, and execution.

John Riggs of Buffalo, New York, has designed and constructed the attractive observing chair shown in Figure B.2. Though

Figure B.2
John Riggs and his rotating observing chair. Note the placement of the platform's pivot points relative to the observer's neck. Photograph supplied by John Riggs.

made for a 3-inch f/5 richest field refractor, the chair is easily adapted to a pair of binoculars. Riggs's inspiration for the chair came from Leslie Peltier's "Merry-Go-Round" observatory and Edgar Everhart's comet seeker, two outstanding designs from the 1940s and 1960s, respectively, as well as from the currently popular Dobsonian telescope mount.

There are four main components to the Riggs observing chair: the seat, seat box, base, and instrument platform. The seat is constructed of ½-inch-thick plywood and hardwood dowels. Since it needs to be custom fit to the individual, this part must be made first. The seat pads and headrest are important factors to consider when building the chair. Four-inch-thick foam is used for the pads, while the cylindrical headrest is made by rolling a thinner, flat piece of foam around a hardwood dowel and then fitting it into a cloth cover.

The seat box is also fabricated from ½-inch-thick plywood. In order to reduce vibration, the seat back never actually touches the seat box. The bottom is secured into the seat box's frame by two machine screws and wing nuts. These permit easy disassembly for transporting the chair to rural observing sites.

A frame of 2×4's and ¾-inch plywood comprises the base of the chair. Note the hexagonal shape, which is needed in order to provide enough foot room for the observer.

The seat–seat box assembly and the base are connected by a large, centrally located bolt. Originally, a sheet of Formica was glued to the bottom of the seat box, and three equally spaced Teflon pads were tacked onto the top of the base, like a Dobsonian telescope mount. The Teflon and Formica combined for a smooth and sturdy, yet inexpensive, bearing. More recently, a ball-bearing Lazy Susan has been substituted, reducing the force needed to rotate the chair even further.

The ¾-inch plywood sides of the instrument platform are affixed to maple cross members to which the instrument is mounted. The altitude bearings are short sections of three-inch-diameter aluminum tubing. The bearings ride on two Teflon pads inside each of the semicircular cutouts on the seat box. Note that, for added rigidity, these cutouts are reinforced with a second piece of plywood. To counterbalance the mass of the instrument, two adjustable steel weights are attached to the ends of the platform arms.

Riggs notes that his observing chair, which was constructed with only simple hand tools, is "the ultimate" for viewing the sky. He uses it for monitoring variable stars, comet seeking, and just enjoying the universe.

While the previously described mountings work fine for most binoculars, they are not strong enough to support truly giant glasses. This was the problem faced by Pearson Menoher of Greenwich, Connecticut. He found that heavy-duty tripods proved too awkward, so he began to build several prototype devices to hold his 60-pound Fujinon 25×105 binoculars.

Menoher's early attempts evolved into the enclosed steel-framed observing chair shown in Figure B.3. The design is built around a surplus airport cargo transporter. Resting on concrete blocks, the transporter frame serves as both a rock-steady foundation and an azimuth bearing. The latter function makes good use of the transporter's four-inch-wide, five-foot-diameter steel track.

The combined weight of the chair, frame, shelter, and observer is carried on the steel track by three steel wheels and one rubber drive wheel. A sleeve bearing, centrally located on the frame, keeps the wheels from derailing. The rubber wheel is driven by a reversible $\frac{1}{3}$-horsepower AC motor. Found inside the shelter, the motor has a reduction of 28,500-to-1 and produces a shelter rotation time of 16.3 minutes.

The chair itself is a welded frame of steel angles with wooden cross members. Seat pads taken from an old easy chair provide comfort and support for the observer. The all-important headrest can be adjusted to the individual, as can be the footrest, taken from an old barber's chair.

Unlike the Riggs chair, where only the instrument platform swings in altitude, the entire Menoher chair pivots in elevation with the platform affixed to it. To move in altitude, a second $\frac{1}{3}$-horsepower electric motor turns a long $\frac{3}{4}$-inch-diameter shaft threaded to a coupler on the chair's footrest. Pivoting on two side bearings, the chair can travel from a near-horizontal position to the zenith in three minutes.

Control switches for both the altitude and azimuth motors are found at the ends of the armrests. Their convenient location

Figure B.3
Pearson Menoher at the controls of his binocular chair observatory.
Photograph by the author.

permits the user to aim the binoculars at any point in the sky without ever having to gaze away from the eyepieces. Limiting switches guarantee that the observer will not accidentally tilt the chair beyond safe limits.

The walls of the shelter are framed with two-inch wooden members and covered with ¼-inch exterior grade plywood. These give excellent protection from the wind, while also sealing out the elements. Wide shelves found along the inside walls are used for charts, a radio, and lights.

Menoher spends most of his observing time monitoring variable stars for the American Association of Variable Star Observers. The slow, smooth altitude and azimuth motions also make this an ideal instrument for comet hunting.

Norman Butler is the proud owner of the beautiful homemade pair of 6-inch binoculars shown in Figure B.4. The finished product took this San Diego, California, amateur nearly three years to design and build. Surely you will agree that his patience paid off nicely.

The optical system is composed of two identical 6-inch f/15 Cassegrain telescopes. For binoculars to function properly, both sides must be the same focal length; otherwise, there would be noticeable differences in effective magnification. When the pair of primary mirrors and matching 2.36-inch Cassegrain secondary mirrors were ordered, the manufacturer was told that each set had to match the other closely in focal length. Happily, after testing the mirrors on an optical bench, no noticeable disparities were found. Instead of using prisms to reflect light exiting the telescope tubes to the eyepieces, Mr. Butler chose optically flat first-surface mirrors. The image bounces off a 1½-inch mirror to a pair of 1¼-inch diagonal mirrors and out through the eyepieces. The eyepieces themselves are taken from a pair of zoom binoculars. These give a magnification range of 105× to 315×, which proves

Figure B.4
Norman Butler proudly displaying his exquisite home-built 6-inch Cassegrain binocular telescope. Note the fine workmanship of the mounting. Photograph supplied by Norman Butler.

quite adequate for viewing nearly all types of celestial objects.

The twin 6-inch telescope assemblies are mounted together in parallel and ride between a pair of 17¾-inch-diameter steel rings. Four small pulleys support the binoculars against the inner rim of each ring, to permit the instrument to rotate as a unit. This alleviates any problem in eyepiece accessibility that may arise at odd viewing angles. The pulleys are attached to adjustable brackets, which are bolted onto the central mating tube.

The binoculars-ring assembly is held on a nicely executed home-built fork mount. A commercially purchased clock drive, attached to the right ascension axis, tracks the instrument with the stars.

There are five mirrors to align on each side of the binoculars, making optical collimation a painstaking task. Only after a time-consuming three-step process of adjusting one primary-secondary mirror pair, then the other, and finally the common diagonal mirror assembly, can correct alignment be achieved. With proper collimation, the two images remain perfectly merged even at the highest magnification.

The stereo-optical effect of these giant binoculars provide dramatic views of the lunar landscape, the planets, and deep sky objects. Butler has glimpsed 14th-magnitude planetary nebulae through his binoculars on good nights, further testimonial to the improved performance of binocular vision over monocular viewing. (Typically, a 6-inch single-eyepiece telescope can penetrate only to 13th magnitude.) Friends have commented that looking through this outstanding instrument is like peering out through a porthole of a spaceship!

Figure B.5
Lee Cain and his mammoth 17½-inch binocular telescope. Photograph supplied by Lee Cain.

While most of us think of giant binoculars as perhaps 10×70's up to about 25×105's, Lee Cain has a different idea. An amateur astronomer and telescope maker from Houston, Texas, Cain has conceived and built three truly huge binoculars, with apertures of 10, 13, and 17½-inches!

Cain's inspiration comes from the same studies referred to in Chapter 1 that show a gain of up to 40% in the visibility of low-contrast objects. As proof of this, observers peering through his binoculars constantly rave about the contrast and apparent brightness of extended nebulae and galaxies, as well as the extraordinary detail visible.

All of Cain's binoculars are twin Newtonian reflecting telescopes by design. Of the three, he is most famous for his 17½-inch f/4.5 binocular telescope (Figure B.5). Based

on "hardware-store technology," as he calls it, the instrument required only hand tools to construct. To ensure the lightest weight and simplest design possible, Cain utilized an open-truss tube and a Dobsonian altitude-azimuth mount, two popular trends among amateur telescope makers.

The binoculars' framework is fabricated from exterior-grade plywood. The base and mirror box were glued and screwed together and then laminated with Formica for improved appearance and some added stiffness. Due to the instrument's weight, Cain chose rubber casters over the more typical Teflon-on-Formica bearing surface for the Dobsonian mount's azimuth axis. Still, the casters compress slightly, giving a "mushy" feeling. He suggests that steel wheels augmented by adjustable friction pads might be a better choice. The altitude bearings ride on pads of Teflon, as on standard Dobsonians.

The primary mirrors are supported by a modified sling system. Using no compressible material such as felt or leather to brace them, the mirrors hold better collimation than more traditional sling designs. This is an especially important consideration in a binocular telescope.

The secondary-tertiary mirror assembly was constructed from double-wall cardboard tubing used as concrete column forms in building construction. The forward end of each telescope rotates on Teflon pads for adjustment of the interpupillary distance. Cain notes that this adjustment is not as critical as might be thought at first. In fact, of all the people who have looked through his instrument, only a few have required a change in eyepiece spacing.

The views through Cain's binoculars are nothing short of phenomenal. Bright objects, such as the Orion Nebula, Lagoon Nebula, and globular cluster Omega Centauri, offer stunning detail that rivals, or even surpasses, the best photographs. Fainter targets, such as the Veil Nebula, Pelican Nebula, and Horsehead Nebula, are not just tentatively visible (as they may be in a 17½-inch monocular telescope), they are easily seen and intricately detailed. Cain's 17½-inch binoculars perhaps prove the oft-heard statement that they do things in a big way down in Texas!

Appendix C

Care, Maintenance, and Other Tidbits

Binoculars are precise optical instruments that require care and occasional maintenance in order to ensure their continued optimal usefulness. Many of the objects discussed in this book may be at the very limits of your binoculars' capabilities of detection and resolution. If your glasses are not in peak operating condition, these objects will remain invisible.

As they say, the best offense is a good defense. This is especially true when it comes to the care of optics. The best way to fight dirt and dust is simply to keep the lens caps on whenever the binoculars are not in use. Even if the glasses are in their case, dust will find its way onto uncapped lenses.

Perhaps your binoculars are older and their original lens caps have been long since lost. Steps can still be taken to seal out dust and dirt. Simply place the glasses into a clean plastic storage bag, such as the kind used in the kitchen to keep leftovers. Evacuate as much air out of the bag as possible before sealing it with a twist tie. Then place the bagged binoculars in their case.

Never clean the lenses just for the sake of cleaning them; dust and dirt should only be dealt with when they become apparent to the eye. Sealing the glasses will go a long way toward preventing this, but when the inevitable happens and a lens becomes contaminated, it is best to take action as soon as possible. Scratches and/or damage to the lens coating may easily result if contaminants are left unattended. However, before any cleaning is attempted, remember that both glass and optical coatings are relatively soft, while dirt and dust may be considered hard. If cleaning is not done correctly, then the abrasive dust can scratch the vulnerable optical elements. Remember, dirt can be cleaned, but scratches are forever, so *be careful!*

The intent of this section is to discuss the cleaning of the lenses' outer surfaces only. Under no circumstances should you dismantle the binoculars. Dirt and dust should never enter the glasses, as binoculars are sealed systems. If, for some reason, the prisms or interior lens surfaces become contaminated, they should be disassembled and cleaned only by a qualified professional. Otherwise, misalignment and other damage may result.

Begin the exterior lens-cleaning process by removing all abrasive particles. This is most easily and safely done with either a soft brush or a can of compressed air. Both are available from photographic supply stores.

If the brush is your choice, be certain to use one made specifically for the purpose. Camel's-hair brushes are preferred. Lightly sweep the surface of the lens in one direction only, flicking the brush free of any accumulated dust particles at the end of each stroke.

Personally, I prefer using a can of compressed air instead of a brush for dusting optical surfaces. That way, the lens is never actually touched. Hold the nozzle of the can away from the lens *at least* as far as recommended by the manufacturer; if the can is held too close, there is a chance that some of the spray propellant will strike the glass surface and stain it. Several short bursts of air are preferred to one long blast.

Once the dust is removed, a gentle detergent may be used for stains and other residue. In a gallon container, mix three quarts of distilled water with one quart of pure-grade isopropyl alcohol, and add two or three drops of liquid dishwashing detergent. Dampen a piece of sterile surgical cotton with the solution, making sure that the cotton is only damp and not dripping. Gently wipe the lens in a circular motion with the moist cotton. Once done, use a second, dry piece of cotton to blot up any moisture on the lens. Repeat this procedure for the other objective lens and eyepieces, but only if needed. Finally, let the binoculars air dry by placing them on a clean horizontal surface.

Many readers can probably recall nights when sky conditions were exceptional. On such nights, you aim your binoculars excitedly toward the heavens. The star-filled vault is so fantastic that you can hardly believe it is real. But after a while,

the stars seem to be slowly disappearing. Is it clouding up? You pull away from the eyepieces to see that the sky is still crystal clear. What's going on here?

The answer is that, although the sky is clear, the binoculars have fallen victim to an extremely localized cloud cover: the objective lenses have fogged over! When humidity is high or the binoculars are colder than the surrounding air, fog will form on the front of the objectives.

Fogging is easily combatted by placing a dew cap around each objective. Ideally, the dew caps should be at least twice as long as the objectives' diameter. The interior must be painted flat black or lined with black paper or felt to dampen any stray reflections.

Soup cans are fine as dew caps for smaller glasses. Line one end with self-adhesive foam weather stripping, which is available from any hardware store. The foam will provide a secure fit while preventing the caps from scratching the lens barrels.

For larger binoculars, aluminum flashing may be used. First, measure the outer circumference of the lens barrels. Add about 10% for a safety factor, plus an extra inch for overlap. Cut a piece of flashing to this length. Wrap the ends loosely around so that one overlaps the other by the extra inch. Tape the cylinder together using plastic packing tape.

Once the inside is painted black, add enough weather stripping so that each dew cap fits snugly around its objective lens barrel. Dew caps should be removed when the binoculars are not in use and replaced whenever dampness is present.

After lens fogging, nothing can ruin an observing session quicker than having to put up with a street light or a neighbor's porch light. The distraction caused by glare seen out of the corners of your eyes can be enough to cause a faint celestial object to be missed.

To help shield our eyes from extraneous light, most binoculars sold today come with collapsible rubber eyecups. However, while they prove adequate under most conditions, eyecups may not block out all peripheral light. If this is the case, what should you do? Go back inside and watch television? Hardly!

If eyecups alone prove inadequate, try this trick. Take ordinary rubber underwater goggles, cut out and discard the front "window," and then spray them with flat black paint. To test your creation, put the goggles on (after the paint has dried, please!) and go out under the stars. The blackened goggles should provide enough added baffling to keep stray light from creeping around the eyecups' edges and into your eyes. Admittedly, they may make you feel a little like a World War I flying ace, but they should do the job.

These homemade goggles are quite effective against nearby individual lights, but they offer no help against sky glow. Sky glow comes from more distant lights that collectively light up the horizon and shine into the sky. This is the most destructive type of light pollution; it can turn an apparently clear blue daytime sky into a yellowish, hazy night sky of limited usefulness.

Although modern technology has caused the problem in the first place, it also offers a partial solution. Many sources of interference radiate in very narrow regions of the visible spectrum. Accordingly, *light pollution filters* are available that block out these specific wavelengths while allowing all others to pass through with little loss.

Most light pollution filters are made with internally threaded telescope eyepieces in mind and are difficult to attach securely to the unthreaded barrels of binocular eyepieces. Fortunately, a few companies make filters specifically for binoculars. You should contact the manufacturer to see whether its filters will work with your particular binoculars *before* making a purchase.

Keep in mind that light pollution filters work for only a few types of celestial objects. The visibility of nebulae suffers most from sky glow, but benefits most from using the filters. Nebulae that have long been considered impossible targets for the visual astronomer can now be seen with comparative ease. On the other hand, stars, clusters, and galaxies experience little or no enhancement. While filters can never make up for a clear, dark observing site far from city lights, smoke, and haze, they offer urban and suburban observers a chance to see certain deep sky objects that were once reserved for rural amateurs only.

Appendix D
Binocular Manufacturers

Following is a listing of major manufacturers of binoculars:

Brunton/Lakota, 620 East Monroe Avenue, Riverton, Wyoming 82501 (307) 856–6559.

Bushnell, Division of Bausch and Lomb, 2828 East Foothill Road, Pasadena, California 91107 (213) 577–1500.

Carl Zeiss, Inc., One Zeiss Drive, Thornwood, New York 10594 (914) 681–7603.

Celestron International, 2835 Columbia Street, Torrance, California 90503 (800) 421–1526.

Company Seven Astro-Optics, 14300 Cherry Lane Court, Laurel, Maryland 20707 (301) 953–2000.

Edmund Scientific, 100 East Gloucester Pike, Barrington, New Jersey 08007 (609) 547–3488.

Fujinon, 10 High Point Drive, Wayne, New Jersey 07470 (201) 633–5600.

Jason Empire, Inc., 9200 Cody, Box 14930, Overland Park, Kansas 66214 (913) 888–0220.

Jena Scientific Instruments, 820 Second Avenue, New York, New York 10017 (212) 867–3051.

Leica USA, Incorporated, 156 Ludlow Street, Northvale, New Jersey 07647 (201) 767–7500.

Meade Instruments Corporation, 1675 Toronto Way, Costa Mesa, California 92626 (800) 854–7481.

Minolta Corporation, 101 Williams Drive, Dept. BIN, Ramsey, New Jersey 07446 (201) 825–4000.

Nikon, Incorporated, 623 Stewart Avenue, Garden City, New York 11530 (516) 222–0200.

Orion Telescope Center, P.O. Box 1158, Santa Cruz, California 95061 (800) 447–1001.

Parks Optical, 270 Easy Street, Simi Valley, California 93065   (800) 447–2757.

Pentax Corporation, 35 Inverness Drive East, Englewood, Colorado 80112   (303) 799–8000.

Selsi Company, Incorporated, 40 Veterans Blvd., Carlstadt, New Jersey 07072   (201) 935–0388.

Swarovski Optik, 2 Slater Road, Cranston, Rhode Island 02920   (401) 463–3000.

Swift Instruments, Incorporated, 952 Dorcester Avenue, Boston, Massachusetts 02125   (617) 436–2960.

Tasco, 7600 NW 26th Street, Miami, Florida 33122   (305) 591–3670.

Unitron Company, Incorporated, 175 Express Street, Plainview, New York 11803 (516) 822–4601.

Appendix E
Converting Universal Time to Local Time

In order to standardize when a particular celestial event is due to occur, astronomers have chosen to express the time of day according to "Universal Time," or simply, U.T. Universal Time (also called Greenwich Mean Time or Coordinated Universal Time) is based on the clock time at the Greenwich, England, prime meridian. To find out when a celestial event is to take place from your location, its U.T. must be converted to local clock time. In general, though not always, there is an hour's difference for every 15° of longitude traveled away from the Earth's prime meridian. Table E.1 shows how to convert Universal Time to various time zones around the world. Remember to

TABLE E.1 Converting Universal Time to Local Time

Time Zone	Local Standard Time*
North America	
Atlantic Standard Time	U.T. − 4 hours
Eastern Standard Time	U.T. − 5 hours
Central Standard Time	U.T. − 6 hours
Mountain Standard Time	U.T. − 7 hours
Pacific Standard Time	U.T. − 8 hours
Yukon Standard Time	U.T. − 9 hours
Alaska-Hawaii Standard Time	U.T. − 10 hours
Great Britain	U.T.
Europe	U.T. + 1 hour
Australia (west coast)	U.T. + 8 hours
Australia (east coast)	U.T. + 10 hours
New Zealand	U.T. + 12 hours

*Note: If you are converting to daylight saving time, be sure to *add* an extra hour (or subtract an hour less, depending how you look at it) to the local time.

change both the time *and* date if necessary.

As an example, suppose a lunar eclipse is due to begin at 07:20 U.T. on a Wednesday. Then an observer in the Eastern Time Zone of the United States would see the eclipse start at 2:20 a.m. Wednesday morning, while the event would commence at 11:20 p.m. Tuesday evening on the Pacific coast. That same eclipse would begin at 7:20 p.m. Wednesday evening for New Zealanders.

Appendix F
For Further Information . . .

The following society addresses and telephone "hot lines" are of general astronomical interest:

Societies

Amateur Satellite Observers, HCR 65, Box 261-B, Kingston, Arkansas 72742

American Association of Variable Star Observers, 25 Birch Street, Cambridge, Massachusetts 02138

American Lunar Society, P.O. Box 209, East Pittsburgh, Pennsylvania 15112

American Meteor Society, Department of Physics, SUNY Geneseo, Geneseo, New York 14454

Association of Lunar and Planetary Observers, P.O. Box 143, Heber Springs, Arkansas 72543

Astronomical League, 6235 Omie Circle, Pensacola, Florida 32504

Astronomical Society of the Pacific, 390 Ashton Avenue, San Francisco, California 94112

Aurora Alert Hotline, Manley Drive, RR#1, Box 232A, Pascoag, Rhode Island 02859

International Dark-Sky Association, 3545 North Stewart, Tucson, Arizona 87516

International Occultation Timing Association, 1177 Collins, Topeka, Kansas 66604

National Deep-Sky Observers Society, 3123 Radiance Road, Louisville, Kentucky 40220

Small Scope Observers Association, 4 Kingfisher Place, Audubon Park, New Jersey 08106

Webb Society (North American Division), 1440 South Marmora Avenue, Tucson, Arizona 85713

Western Amateur Astronomers, 13617 East Bailey, Whittier, California 90601

Telephone Hot Lines:

Abrams Planetarium, East Lansing, Michigan: (517) 332–7827 (general sky information)

Griffith Planetarium, San Francisco, California: (213) 663–8171 (general sky information)

Maryland Scientific Center, Baltimore, Maryland: (301) 539–7827 (general sky information)

National Air and Space Museum, Washington, D.C.: (202) 357–2000 (general sky information)

Sky and Telescope, Cambridge, Massachusetts: (617) 497–4168 (general sky information and new discoveries)

University of Illinois, Urbana, Illinois: (217) 333–8789 (general sky information)

U.S. Naval Observatory, Washington, D.C.: (900) 410–8463 (official time; 50 cents per minute from anywhere in United States)

WWV, Boulder, Colorado: (303) 499–7111 (precise time signals)

Appendix G
Bibliography

Among the many excellent books and periodicals on astronomy and related topics published today are the following:

Astronomy (monthly periodical). P.O. Box 1612, Waukesha, Wisconsin 53187.

Burnham, R., Jr. *Burnham's Celestial Handbook,* volumes 1–3. Dover, 1978.

Cherrington, E. *Exploring the Moon through Binoculars and Small Telescopes.* Dover, 1983.

Deep Sky (quarterly periodical). P.O. Box 1612, Waukesha, Wisconsin 53187.

Eicher, D., et al. *Deep Sky Observing With Small Telescopes.* Enslow, 1989.

Eicher, D. *Universe from Your Backyard.* Cambridge University Press, 1988.

Espenak, F. *Fifty Year Canon of Lunar Eclipses: 1986—2035.* Sky Publishing, 1988.

Espenak, F. *Fifty Year Canon of Solar Eclipses: 1986—2035.* Sky Publishing, 1989.

Hartung, E. *Astronomical Objects for Southern Telescopes.* Cambridge University Press, 1985.

International Comet Quarterly (quarterly periodical). Smithsonian Astrophysical Observatory, 60 Garden Street, Cambridge, Massachusetts 02138.

Jones, K., et al. *Webb Society Deep-Sky Observer's Handbook*, volumes 1–7. Enslow, 1979–1987.

Levy, D. *Observing Variable Stars.* Cambridge University Press, 1989.

Lorenzin, T., with Sechler, T. *1000+, the Amateur Astronomer's Field Guide to Deep Sky Observing.* Lorenzin, 1987.

Mallas, J., and Kreimer, E. *The Messier Album.* Sky Publishing, 1978.

Meeus, J. *Astronomical Tables of the Sun, Moon and Planets.* Willmann-Bell, 1983.

Menzel, D., and Pasachoff, J. *Field Guide to the Stars and Planets.* Houghton Mifflin, 1983.

Motz, L., and Nathanson, C. *The Constellations: An Enthusiast's Guide to the Night Sky.* Doubleday, 1988.

Newton, J. and Teece, P. *Guide to Amateur Astronomy.* Cambridge University Press, 1988.

Observer's Guide (bimonthly periodical). P.O. Box 35, Natrona Heights, Pennsylvania 15065

Ottewell, G. *Astronomical Calendar* (annual). Furman University.

Paul, H. *Binoculars and All-Purpose Telescopes*. American Photographic Book Publishing, 1980. (Sadly, this book is currently out of print.)

Percy, R., et al. *Observer's Handbook* (annual). Royal Astronomical Society of Canada.

Ridpath, I., et al. *Norton's 2000.0* (star atlas). John Wiley and Sons, 1989.

Scovil, C. *AAVSO Star Atlas*. Sky Publishing, 1980.

Sherrod, C. *Complete Manual of Amateur Astronomy*. Prentice-Hall, 1981.

Sidgwick, J. *Observational Astronomy for Amateurs*. Enslow, 1982.

Sinnott, R. *NGC 2000.0*. Sky Publishing, 1988.

Sinnott, R., and Hirschfield, A. *Sky Catalogue 2000.0*, volumes 1, 2. Sky Publishing, 1985.

Sky and Telescope (monthly periodical). P.O. Box 9111, Belmont, Massachusetts 02178.

Tirion, W. *Sky Atlas 2000.0* (star atlas). Sky Publishing, 1981.

Tirion, W., Rappaport, B., and Lovi, G. *Uranometria 2000.0*, volumes 1, 2 (star atlas). Willmann-Bell, 1987.

Index

The index offers reference to general discussions, definitions, and sky objects. It does not list all deep sky objects in Chapter 7, but rather only those specifically detailed. For a complete listing, review the constellation tables found throughout Chapter 7.

Page numbers with illustrations are followed by *i* (e.g., 6*i*).

Abbreviations, deep sky object, 82
Acrux, 135
Adams, J., 34
AG Draconis, 149
AG Pegasi, 198
Aitken, R., 72
Albategnius (crater), 6*i*, 8
Albireo, 56*i*, 139
Alcock, G., 80
Alcor & Mizar, 249
Alcyone, 237*i*
Algol, 57, 201
Aliacenis (crater), 6*i*, 8
Almagest, 69
Alpha¹⁺² Capricorni, 107–108
Alpha Centauri, 120
Alpha Crucis, 135
Alpha Herculis, 158
Alpha Leonis, 165
Alpha¹⁺² Librae, 169
Alpha Orionis, 194

Alpha Persei Cluster, 201, 202*i*
Alpha Piscis Austrini, 206–207
Alphonsus (crater), 13*i*, 14
Alpine Valley, 13*i*, 18
Alps Mountains, 13*i*, 18
American Association of Variable Star Observers, 76
Anaxagoras (crater), 13*i*, 14
Andromeda, 86–88
Andromeda Galaxy, 86, 87*i*
Antlia, 88
Apennine Mountains, 13*i*, 18
Apus, 88–89
Aquarius, 89–91
Aquila, 91–92
Ara, 92–94
Archimedes (crater), 13*i*, 15
Aries, 94
Aristarchus (crater), 13*i*, 15
Aristillus (crater), 13*i*, 15

Aristoteles (crater), 6*i*, 8
Arzachel (crater), 13*i*, 15
Asteroids, 37–39, (38*i*)
Asterope, 236–237*i*
Astronomy with an Opera Glass, v
Atlas (crater), 6*i*, 9
Auriga, 95–97

Barnard, E., 71
Barnard 33, 64*i*
Barnard 59, 188
Barnard 64, 188
Barnard 65, 188
Barnard 66, 188
Barnard 67, 188
Barnard 72, 188–189, 217*i*
Barnard 78, 188
Barnard 88, 219
Barnard 89, 219
Barnard 92, 218
Barnard 103, 232
Barnard 111, 233

Barnard 112, 233
Barnard 119a, 233
Barnard 133, 92
Barnard 142–3, 92
Barnard 168, 144
Barnard 228, 171
Barnard 296, 219
Barnard 318, 233
Barnard's Galaxy, 222
Barnard's Star, 189
Bayer, J., 69
BB Sagittarii, 221
Beehive Cluster, 38i, 101
Bernes 157, 131
Beta^{1+2} Capricorni, 108
Beta Cygni, 56i, 139
Beta Lyrae, 782i, 173
Beta Persei, 57, 201
Betelgeuse, 194
Binary Stars, (see Stars,
      double and multiple)
Binoculars
   choosing, 264–268
   comparisons of, 266
   exit pupil, 265
   German-type v. American-
      type, 264
   giant, 264
   porro prism v. roof prism,
      263–264
   purchasing, 266–268
   supporting, 269–275
   wide angle, 265–266
   zoom, 267
Binoculars and All-Purpose
   Telescopes, 267
Black-Eye Galaxy, 130
Blancanus (crater), 13i, 15
Blanco 1, 228
Blinking Planetary, 139
BM Scorpii, 226
Bochum 10, 112
Bochum 11, 112
Boötes, 97–98
Brocchi's Cluster 258-259i
BU Geminorum, 155
Bullialdus (crater), 13i, 15
Burnham 584, 101
Burns, J., 270
Butler, N., 273–274

Butterfly Cluster, 226, 227i
Byrgius A (crater), 13i, 15

Caelum, 98
Cain, L., 274–275
California Nebula, 202, 203i
Camelopardalis, 99–100
Cancer, 100–101
Canes Venatici, 102–104
Canis Major, 104–106
Canis Minor, 107
Capricornus, 107–108
Carina, 108–112
Carpathian Mountains, 13i,
      19
Cassiopeia, 113–115
Catalogue of 349 Dark
   Objects in the Sky, 71
Catalogues,
   cross-reference index, 83–
      84
   deep sky object, 68–72
Caucasus Mountains, 6i, 12
Centaurus, 115–120
Centaurus A, 117–118i
Cepheid variable, 59, 123
Cepheus, 121–123
Cepheus OB2 Association,
      123
Cetus, 123–125
Chamaeleon, 125–126
Chi Cygni, 139
Chi Persei, 199–200i
Christmas Tree Cluster, 180
Circinus, 126
Clavius (crater), 13i, 15
Cleomedes (crater), 6i, 9
Clusters, Star, (see Star
   clusters)
Coalsack, 135–136i
Coathanger Cluster, 258–259i
Cocoon Nebula, 144
Collinder 21, 242
Collinder 62, 95–96
Collinder 65, 191
Collinder 69, 192
Collinder 70, 194,195i
Collinder 89, 155
Collinder 91, 178
Collinder 96, 178

Collinder 97, 178
Collinder 106, 180
Collinder 107, 180
Collinder 121, 105–106
Collinder 135, 208
Collinder 140, 106
Collinder 228, 112
Collinder 240, 112
Collinder 285, 247–248
Collinder 299, 184
Collinder 316, 225
Collinder 350, 189
Collinder 394, 221
Collinder 399, 258–259i
Collinder 464, 100
Columba, 126–127
Coma Berenices, 127–130
Coma Berenices Star Cluster,
      128, 129i
Comets, 39–44, (40i)
   estimating diameter, 41
   estimating magnitude, 40–
      41
   estimating tail length, 41
   hunting, 42–44, (43i)
   position angle, 42
   reporting new, 44
Copernicus (crater), 13i, 16
Corona Australis, 130–131
Corona Borealis, 131–132
Corvus, 133
Crab Nebula, 66i, 238–239
Crater, 133
Crux, 134–137
Cygnus, 137–144
Cyrillus (crater), 6i, 9

Delphinus, 144–145
Delta^{1+2} Apodis, 89
Delta Centauri, 117
Delta Cephei, 59, 123
Delta^{1+2} Chamaeleonis, 125–
      126
Delta^{2} Gruis, 157
Delta^{1+2} Lyrae, 174
Delta Orionis, 191
Delta^{1+2} Telescopii, 240
Dew caps, constructing, 277
Diamond-Ring asterism, 250
Dolidze 26, 107

Dolidze-Dzimselejsvili, 4, 239
Dolidze-Dzimselejsvili, 9, 159
Dorado, 145–148
Double Cluster, 199–200i
Double stars, (see Stars, double and multiple),
Double-Double, 173
Draco, 148–150
Dreyer, J., 71
Dumbbell Nebula, 260, 261i
Dunlop 236, 176

Eagle Nebula, 63i, 234
Eclipses, lunar, 21–24, (22i, table 24)
Eclipses, solar, 51–54, (52i, 53i, table 54)
Eight-Burst Nebula, 253
Endymion (crater), 6i, 9
Epsilon¹⁺² Lyrae, 173
Equuleus, 150
Eratosthenes (crater), 13i, 16
Eridanus, 150–151
Eskimo Nebula, 155–156
Eta Carinae Nebula, 111i–112
Eta Tauri, 237i
Exit Pupil, (see Binoculars, exit pupil)

Fabricius (crater), 6i, 9
Feinstein 1, 112
54 Sagittarii, 222
Filters, light pollution, 278
5 Lyncis, 172
Fomalhaut, 206–207
Fornax, 152–153
42 & 45 Orionis, 193i, 194
47 Tucanae, 243–244, 245i
Fracastorius (crater), 6i, 9
Furnerius (crater), 6i, 9

Galaxies (see also specific entries), 65–69
  elliptical, 68i
  irregular, 68, 69i
  spiral, 66–67i
Galileo, 30
Galle, J., 34
Gamma Cassiopeiae, 60, 114

Gamma Circini, 126
Gamma Crucis, 135
Gamma Equulei, 150
Gamma Leporis, 168
Gamma Velorum, 252
Gassendi (crater), 13i, 16
Gemini, 153–156
General Catalogue of Nebulae, 70–71
Ghost of Jupiter Nebula, 161
Goggles, light pollution, 278
Grimaldi (crater), 13i, 16
Grus, 156–157

h 3780, 168
h 5114, 240
h Persei, 199–200i
H V 38, 132
Haemus Mountains, 6i, 12
Haffner 13, 209
Harrington catalogue, 71
Harrington 1, 250
Harrington 2, 152–153
Harrington 3, 99
Harrington 4, 95,96i
Harrington 5, 180
Harrington 6, 167
Harrington 7, 158
Harrington 8, 240
Harrington 9, 145
Harrington 10, 141
Harrington 11, 123
Harrington 12, 115
Harvard 10, 184
Harvard 20, 213
HD 153919, 226
Helix Nebula, 90, 91i
Hercules, 157–159
Hercules (crater), 6i, 9
Herschel's Garnet Star, 122–123
Herschel, J., 70–71
Herschel, W., 32, 70
Hevelius (crater), 13i, 16
Hipparchus, 69
Hipparchus (crater), 6i, 10
Horologium, 159
Horsehead Nebula, 64i
Hubble E., 66

Hubble's Variable Nebula, 180
Huygens, C., 32
Hyades, 238
Hydra, 160–162
Hydrus, 162

IC 1396, 122
IC 2391, 252–253
IC 2395, 253
IC 2488, 253
IC 2581, 110
IC 2602, 110–111
IC 2944, 117
IC 2948, 117
IC 4651, 93–94
IC 4665, 189
IC 4725, 219–220
IC 4756, 234
IC 5067/70, 141,142i
IC 5146, 144
Index Catalogue (IC), 71
Indus, 163
Iota Boötis, 98
Iota Cancri, 101
Iota Librae, 169–170
Iota Microscopii, 176

Janssen (crater), 6i, 10
Jewel Box Cluster, 136–137
Jupiter, 30–31i

Kappa Pavonis, 196
Kappa¹⁺² Volantis, 257
Kepler (crater), 13i, 16

L² Puppis, 208
Lacerta, 163–164
Lacus Mortis, 6i, 12
Lacus Somniorum, 6i, 12
Lagoon Nebula, 216, 218, 227i
Lambda Arietis, 94
Lambda Pavonis, 196
Langrenus (crater), 6i, 10
Large Magellanic Cloud, 146, 147i
Lens cleaning solution, formula for, 277
Leo, 164–166

Leo Minor, 167
Lepus, 167–168
Leverrier, U., 34
Libra, 169–170
Little Man Cluster, 114–115
Longomontanus (crater), 13*i*, 16
Lupus, 170–171
Lynds 906, 140
Lynds 935, 141,142*i*
Lynx, 171–172
Lyra, 172–175

M1, 66*i*, 238–239
M2, 90
M3, 103–104
M4, 224–225
M5, 234
M6, 226, 227*i*
M7, 226, 227*i*, 228
M8, 216, 218, 227*i*
M9, 188
M10, 187
M11, 232*i*–233
M12, 187
M13, 62*i*, 158
M14, 189
M15, 197–198
M16, 63*i*, 234
M17, 219
M18, 219
M19, 188
M20, 216
M21, 218
M22, 220*i*, 221
M23, 216, 217*i*
M24, 218
M25, 219–220
M26, 232
M27, 260, 261*i*
M28, 219
M29, 140
M30, 108
M31, 86,87*i*
M32, 86–87*i*
M33, 240–241*i*
M34, 201
M35, 154*i*–155
M36, 96*i*
M37, 96*i*, 97

M38, 96*i*
M39, 143*i*–144
M40, 248–249
M41, 105, 106*i*
M42, 63, 192–194 (193*i*, 195*i*)
M43, 192, 193*i*
M44, 38*i*, 101
M45, 236–237*i*
M46, 209
M47, 209
M48, 160–161
M49, 255
M50, 181
M51, 103, 104*i*
M52, 115
M53, 130
M54, 221
M55, 222
M56, 174
M57, 174
M58, 256
M59, 257
M60, 257
M61, 255
M62, 187
M63, 103
M64, 130
M65, 165–166*i*
M66, 165–166*i*
M67, 61*i*, 101
M68, 161
M69, 219
M70, 221
M71, 214
M72, 89–90
M73, 90
M75, 222–223
M77, 125
M78, 194
M79, 168
M80, 224
M81, 67*i*, 246–247
M82, 69*i*, 246–247
M83, 162
M84, 68*i*, 255
M85, 128–129
M86, 68*i*, 255
M87, 256
M88, 129

M89, 256
M90, 256
M92, 158–159
M93, 210
M94, 103
M95, 165
M96, 165
M99, 128
M100, 128
M101, 249
M103, 114
M104, 256–257
M105, 165
M106, 102
M107, 187
M108, 247
M110, 87*i*
Manzinus (crater), 6*i*, 10
Mare Cognitum, 12, 13*i*, 14
Mare Crisium, 6*i*, 7
Mare Fecunditatis, 6*i*, 7
Mare Frigoris, 6*i*, 7
Mare Humorum, 13*i*, 14
Mare Imbrium, 13*i*, 14
Mare Marginis, 6*i*, 7
Mare Nectaris, 6*i*, 7
Mare Nubium, 13*i*, 14
Mare Serenitatis, 6*i*, 7–8
Mare Smythii, 6*i*, 8
Mare Tranquillitatis, 6*i*, 8
Mare Vaporum, 6*i*, 8
Mars, 29*i*–30
Maurolycus (crater), 6*i*, 10
Melotte 20, 201, 202*i*
Melotte 25, 238
Melotte 101, 111
Melotte 111, 128, 129*i*
Melotte 186, 189–190
Melotte 227, 185
Menoher, P., 272–273
Mensa, 175
Mercury, 26–27
Messier, C., 69–70
Messier Catalogue, 70, 73–75
Messier Marathon, 73, 75
Meteors, 44–46, (45*i*)
Metius (crater), 6*i*, 10
Microscopium, 176
Mira, 58*i*, 77*i*, 124–125
Monoceros, 177–180

Moon, 3–24, (4i, 6i, 13i, 20i, 22i)
  eclipses, 21–24, (22i, table 24)
  map, First Quarter, 6i
  map, Last Quarter, 13i
  occultations, 20i–21
  surface features (see also specific features), 6i–19, (13i)
  Young/Old, 4i–5, 27i
Moretus (crater), 13i, 17
Morris, C., 40
Mountings, binocular
  binocular observatory, 272–273
  Cassegrain binoculars, 273–274
  Dobsonian binoculars, 274–275
  flexible parallelogram, 270
  observing chair, 270–272
  tripods, 269
Mu Boötis, 98
Mu Cephei, 122–123
Multiple Stars, (see Stars, double and multiple)
Musca, 181–183
Mutus, 6i, 10

Nebulae (see also specific entries), 62–65
  dark, 64i
  diffuse, 63i
  emission, 63
  planetary, 64–65i
  reflection, 63
  supernovae remnants, 65, 66i
Neptune, 34–35, (33i)
New General Catalogue of Double Stars (ADS), 72
New General Catalogue of Nebulae and Clusters (NGC), 71
NGC 55, 229
NGC 104, 243–244, 245i
NGC 129, 114
NGC 188, 121–122
NGC 205, 87i

NGC 221, 86–87i
NGC 224, 86, 87i
NGC 246, 124
NGC 247, 124
NGC 253, 229i
NGC 288, 229–230
NGC 292, 244, 245i
NGC 346, 244, 245i
NGC 362, 244, 245i
NGC 371, 244, 245i
NGC 457, 114
NGC 581, 114
NGC 598, 240–241i
NGC 602, 162
NGC 604, 242
NGC 752, 87
NGC 869, 199–200i
NGC 884, 199–200i
NGC 1027, 115
NGC 1039, 201
NGC 1068, 125
NGC 1261, 159
NGC 1291, 151
NGC 1316, 152
NGC 1360, 153
NGC 1365, 153
NGC 1435, 237i
NGC 1499, 202, 203i
NGC 1501, 65i
NGC 1502, 99–100
NGC 1528, 203
NGC 1535, 151
NGC 1545, 203
NGC 1647, 238
NGC 1746, 238
NGC 1763, 146, 147i
NGC 1807, 238
NGC 1817, 238
NGC 1851, 127
NGC 1904, 168
NGC 1910, 146, 147i
NGC 1912, 96i
NGC 1952, 66i, 238–239
NGC 1960, 96i
NGC 1976, 63, 192–194, (193i, 195i)
NGC 1977, 193i, 194
NGC 1981, 192, 193i
NGC 1982, 192, 193i
NGC 2017, 168

NGC 2068, 194
NGC 2070, 147i, 148
NGC 2099, 96i, 97
NGC 2158, 155
NGC 2168, 154i–155
NGC 2169, 194
NGC 2232, 178
NGC 2237, 179i–180
NGC 2244, 179i–180
NGC 2261, 180
NGC 2264, 180
NGC 2287, 105, 106i
NGC 2301, 181
NGC 2323, 181
NGC 2362, 106
NGC 2392, 155–156
NGC 2403, 100
NGC 2422, 209
NGC 2423, 209
NGC 2437, 209
NGC 2439, 209
NGC 2447, 210
NGC 2451, 210
NGC 2477, 210
NGC 2516, 109–110
NGC 2527, 210
NGC 2539, 210
NGC 2546, 210–211
NGC 2547, 252
NGC 2548, 160–161
NGC 2571, 211
NGC 2632, 38i, 101
NGC 2669, 253
NGC 2682, 61i, 101
NGC 2683, 172
NGC 2808, 110
NGC 2903, 164–165
NGC 2976, 247
NGC 3031, 67i, 246–247
NGC 3034, 69i, 246–247
NGC 3114, 110
NGC 3115, 235
NGC 3132, 253
NGC 3201, 253–254
NGC 3228, 254
NGC 3242, 161
NGC 3293, 110
NGC 3324, 110
NGC 3351, 165
NGC 3368, 165

NGC 3372, 111*i*–112
NGC 3379, 165
NGC 3532, 112
NGC 3556, 247
NGC 3572, 112
NGC 3623, 165–166*i*
NGC 3627, 165–166*i*
NGC 3628, 165–166*i*
NGC 3766, 116–117
NGC 3909, 117
NGC 3918, 117
NGC 4052, 134
NGC 4103, 135
NGC 4254, 128
NGC 4258, 102
NGC 4303, 255
NGC 4321, 128
NGC 4349, 135
NGC 4372, 182
NGC 4374, 68*i*, 255
NGC 4382, 128–129
NGC 4406, 68*i*, 255
NGC 4463, 182
NGC 4472, 255
NGC 4486, 256
NGC 4501, 129
NGC 4552, 256
NGC 4565, 129–130
NGC 4569, 256
NGC 4579, 256
NGC 4590, 161
NGC 4594, 256–257
NGC 4609, 135
NGC 4621, 257
NGC 4649, 257
NGC 4736, 103
NGC 4755, 136–137
NGC 4815, 182
NGC 4826, 130
NGC 4833, 182
NGC 5024, 130
NGC 5055, 103
NGC 5128, 117–118*i*
NGC 5139, 118–119*i*
NGC 5189, 182–183
NGC 5194, 103, 104*i*
NGC 5195, 103, 104*i*
NGC 5236, 162
NGC 5272, 103–104
NGC 5281, 119–120

NGC 5299, 120
NGC 5457, 249
NGC 5460, 120
NGC 5466, 98
NGC 5822, 171
NGC 5823, 126
NGC 5897, 170
NGC 5904, 234
NGC 5986, 171
NGC 6025, 243
NGC 6067, 183–184
NGC 6087, 184
NGC 6093, 224
NGC 6121, 224–225
NGC 6124, 225
NGC 6152, 184
NGC 6171, 187
NGC 6188, 93
NGC 6193, 93
NGC 6205, 62*i*, 158
NGC 6208, 93
NGC 6210, 158
NGC 6218, 187
NGC 6231, 225
NGC 6242, 226
NGC 6250, 93
NGC 6254, 187
NGC 6266, 187
NGC 6273, 188
NGC 6281, 226
NGC 6322, 226
NGC 6333, 188
NGC 6341, 158–159
NGC 6397, 94
NGC 6402, 189
NGC 6405, 226, 227*i*
NGC 6475, 226, 227*i*, 228
NGC 6494, 216, 217*i*
NGC 6514, 216
NGC 6523, 216, 218, 227*i*
NGC 6530, 218
NGC 6531, 218
NGC 6541, 130
NGC 6543, 149–150
NGC 6603, 218
NGC 6611, 63*i*, 234
NGC 6613, 219
NGC 6618, 219
NGC 6626, 219
NGC 6633, 190

NGC 6637, 219
NGC 6656, 220*i*, 221
NGC 6681, 221
NGC 6694, 232
NGC 6705, 232*i*–233
NGC 6709, 92
NGC 6715, 221
NGC 6716, 221
NGC 6720, 174
NGC 6726, 131
NGC 6738, 92
NGC 6744, 196
NGC 6752, 196–197
NGC 6774, 221
NGC 6779, 174
NGC 6809, 222
NGC 6820, 260
NGC 6822, 222
NGC 6823, 260
NGC 6826, 139
NGC 6838, 214
NGC 6853, 260, 261*i*
NGC 6864, 222–223
NGC 6871, 139–140
NGC 6913, 140
NGC 6934, 145
NGC 6939, 122
NGC 6940, 260–261
NGC 6946, 122
NGC 6981, 89–90
NGC 6992, 140–141
NGC 6994, 90
NGC 7000, 141, 142*i*
NGC 7009, 90
NGC 7039, 143
NGC 7078, 197–198
NGC 7089, 90
NGC 7092, 143*i*–144
NGC 7099, 108
NGC 7209, 164
NGC 7243, 164
NGC 7293, 90, 91*i*
NGC 7331, 198
NGC 7654, 115
NGC 7662, 87–88
NGC 7789, 115
Norma, 183–184
North American Nebula, 141,
    142*i*
Northern Coalsack, 140

Novae, 59, 79–80
Nu Draconis, 149

OB Associations (see also
    specific entries), 60
Oceanus Procellarum, 13i, 14
Octans, 184–185
Omega Centauri 118-119i
Omega Nebula, 219
Omicron Ceti, 58i, 77i, 124–
    125
Omicron² Eridani, 151
Ophiuchus, 185–190
Optical coatings, 266–267
Orion, 190–195
Orion Belt Cluster, 194, 195i
Orion Nebula, 63, 192–194,
    (193i), 195i
OΣΣ 178, 92

Palus Epidemiarum, 13i, 19
Palus Putredinus, 13i, 19
Paul, H., 267
Pavo, 196–197
Pegasus, 197–198
Pelican Nebula, 141, 142i
Perseus, 198–203
Petavius (crater), 6i, 10
Phoenix, 204
Pi¹⁺² Gruis, 156
Pi¹⁺² Hydri, 162
Pi¹⁺² Octantis, 185
Piazzi, G., 37
Piccolomini (crater), 6i, 10
Pictor, 204–205
Pinwheel Galaxy, 249
Pipe Nebula, 188
Pisces, 205–206
Piscis Austrinus, 206–207
Pismis 4, 252
Pitatus (crater), 13i, 17
Planets (see also specific
    entries), 25–35
Plato (crater), 13i, 17
Pleiades, 236–237i
Pluto, 35
Posidonius (crater) 6i, 10-11
Position angle
    comets, 42
    double stars, 56

Praesepe Cluster, 38i, 101
Proxima Centauri, 120
Psi¹ Piscium, 206
Ptolemy, 69
Ptolemy (crater) 13i, 17, 69
Puppis, 207–211
Purbach, 13i, 17
Putter Cluster, 242
Pythagoras, 13i, 17
Pyxis, 211–212

R Caeli, 98
R Cancri, 100–101
R Centauri, 120
R Chamaeleonis, 125
R Coronae Borealis, 60, 132
R Corvi, 133
R Crucis, 135
R Cygni, 139
R Draconis, 149
R Horologii, 159
R Hydrae, 161–162
R Leonis, 78i, 165
R Leonis Minoris, 167
R Leporis, 77i, 168
R Lyrae, 78i, 174
R Pictoris, 205
R Reticuli, 212–213
R Scuti, 232
Rasalgethi, 158
Regulus, 165
Reiner Gamma, 13i, 19
Reticulum, 212–213
Rheita (crater), 6i, 11
Rheita Valley, 6i, 12
Rho Ophiuchi, 186–187
Riggs, J., 270–272
Ring Nebula, 174
Riphaeus Mountains, 13i, 19
Rosette Nebula 179i-180
Roslund 5, 140
RR Lyrae, 59, 175
RU Crateris, 133
Ruprecht 56, 211
Ruprecht 98, 134
Ruprecht 113, 183
Ruprecht 173, 140

S Antliae, 88
S Coronae Borealis, 131–132

S Doradus, 59, 146
S Indi, 163
S Normae, 184
S Trianguli Australis, 242–
    243
S Ursae Minoris, 250
Sagitta, 213–214
Sagittarius, 214–223
Sailboat Cluster, 167
Sandqvist and Lindroos, 42,
    131
Saturn, 31–32i
Saturn Nebula, 90
Scheiner (crater), 13i, 17
Schickard (crater), 13i, 17–18
Schiller (crater), 13i, 18
Scorpius, 223–228
Sculptor, 228–230
Scutum, 230–233
Serpens, 233–234
Serviss, G., vi
17 Comae Berenices, 128
Sextans, 235
Sinus Iridum, 13i, 19
Sinus Medii, 6i, 12
16 & 17 Draconis, 149
61 Cygni, 141–142
Small Magellanic Cloud, 244,
    245i
Small Sagittarius Star cloud,
    218
Smithsonian Astrophysical
    Observatory, 39–40, 44,
    80
Snake Nebula, 188–189, 217i
Societies, astronomical, 283
Sombrero Galaxy, 256–257
Spindle Galaxy, 235
Star clusters (see also specific
    entries), 60–62
    galactic, 60–61i
    globular, 61–62i
    OB Associations, 60
    open, 60–61i
Stars
    double and multiple, 55–
        57, (56i), 72, 76, 84
    variable, 57–60, (58i), 72,
        76, 77i–78i, 79
Stephenson, 1, 174

Stevinus (crater), 6*i*, 11
Stock 1, 260
Stock 2, 114–115
Stock 10, 96–97
Stock 23, 99
Stofler (crater), 6*i*, 11
Straight Wall, 13*i*, 19
Struve, O., 72
Struve, W., 72
Sun, 47–54, (48*i*)
    eclipses, 49–54, (50*i*, 52*i*,
        53*i*, table 54)

T. Antliae, 88
T Centauri, 119
T Coronae Borealis, 132
T Indi, 163
T Microscopii, 176
T Pyxidis, 211–212
Tarantula Nebula, 147*i*, 148
Taruntius (crater), 6*i*, 11
Tau^{1+2} Lupi, 170–171
Taurus, 235–239
Telephone "hotlines,"
    astronomical, 283–284
Telescopium, 239–240
Temple's Nebula (Merope),
    237*i*
Thales (crater), 6*i*, 11
Theophilus (crater), 6*i*, 11
Theta Apodis, 89
Theta^{1+2} Orionis, 192, 193*i*
Theta Pictoris, 205
Theta^{1+2} Tauri, 238
30 Arietis, 94

36 Ophiuchi, 188
Tombaugh, C., 35
Trapezium, 192, 193*i*
Triangulum, 240–242
Triangulum Australe, 242–
    243
Trifid Nebula, 216
Tripods, selecting, 269–270
Trumpler 2, 200–201
Trumpler 10, 253
Trumpler 14, 112
Trumpler 15, 112
Trumpler 16, 111
Trumpler 18, 112
Trumpler 24, 225–226
Tucana, 243–245
TW Centauri, 119
21 & 22 Tauri, 236–237*i*
27 Hydrae, 161
27 & BU Tauri, 236, 237*i*
Tycho (crater), 13*i*, 18
TZ Mensae, 175

U Coronae Borealis, 132
U Delphini, 145
U Geminorum, 60, 156
U Sagittae, 213
U Sagittarii, 220
U Ursae Minoris, 250
Universal time, converting,
    281–282
Upgren 1, 102–103
Uranometria, 69
Uranus, 32–34, (33*i*)
Ursa Major, 246–249

Ursa Major Moving Cluster,
    247–248
Ursa Minor, 250

V861 Scorpii, 226
van den Bos 1833, 225
Variable Stars, (*see* Stars,
    variable)
Veil Nebula, 140–141
Vela, 250–254
Vendelinus (crater), 6*i*, 11
Venus, 27–29, (28*i*)
Virgo, 254–257
Vision, binocular vs.
    monocular, 1–2
Volans, 257
Vulpecula, 258–261

Walter (crater), 13*i*, 18
Werner (crater), 6*i*, 11
Whirlpool Galaxy, 103, 104*i*
Wild Duck Cluster, 232*i*–233
Winnecki 4, 248–249
WZ Sagittae, 214

X Cygni, 140
X-Marks-the-Spot Cluster,
    240

Zeta & 80 Ursae Majoris,
    249
Zeta Lyrae, 173
Zeta Phoenicis, 204
Zeta Reticuli, 212